T0211916

Introduction to Biomedical Engineering: Biomechanics and Bioelectricity Part II

Synthesis Lectures on Biomedical Engineering

Editor
John D. Enderle, *University of Connecticut*

Bioinstrumentation
John D. Enderle
2006

Fundamentals of Respiratory Sounds and Analysis
Zahra Moussavi
2006

Advanced Probability Theory for Biomedical Engineers
John D. Enderle, David C. Farden, and Daniel J. Krause
2006

Intermediate Probability Theory for Biomedical Engineers
John D. Enderle, David C. Farden, and Daniel J. Krause
2006

Basic Probability Theory for Biomedical Engineers
John D. Enderle, David C. Farden, and Daniel J. Krause
2006

Sensory Organ Replacement and Repair
Gerald E. Miller
2006

Artificial Organs
Gerald E. Miller
2006

Signal Processing of Random Physiological Signals
Charles S. Lessard
2006

Image and Signal Processing for Networked E-Health Applications
Ilias G. Maglogiannis, Kostas Karpouzis, and Manolis Wallace
2006

Introduction to Biomedical Engineering: Biomechanics and Bioelectricity - Part II
Douglas A. Christensen

ISBN: 978-3-031-00510-7   paperback
ISBN: 978-3-031-01638-7   ebook

DOI 10.1007/978-3-031-01638-7

A Publication in the Springer series
SYNTHESIS LECTURES ON BIOMEDICAL ENGINEERING

Lecture #29
Series Editor: John D. Enderle, University of Connecticut

Series ISSN
Synthesis Lectures on Biomedical Engineering
Print 1932-0328   Electronic 1932-0336

Introduction to Biomedical Engineering: Biomechanics and Bioelectricity Part II

Douglas A. Christensen
University of Utah

SYNTHESIS LECTURES ON BIOMEDICAL ENGINEERING #29

ABSTRACT

Intended as an introduction to the field of biomedical engineering, this book covers the topics of biomechanics (Part I) and bioelectricity (Part II). Each chapter emphasizes a fundamental principle or law, such as Darcy's Law, Poiseuille's Law, Hooke's Law, Starling's Law, levers and work in the area of fluid, solid, and cardiovascular biomechanics. In addition, electrical laws and analysis tools are introduced, including Ohm's Law, Kirchhoff's Laws, Coulomb's Law, capacitors and the fluid/electrical analogy. Culminating the electrical portion are chapters covering Nernst and membrane potentials and Fourier transforms. Examples are solved throughout the book and problems with answers are given at the end of each chapter. A semester-long Major Project that models the human systemic cardiovascular system, utilizing both a Matlab numerical simulation and an electrical analog circuit, ties many of the book's concepts together.

KEYWORDS

biomedical engineering, biomechanics, cardiovascular, bioelectricity, modeling, Matlab

To Laraine

Contents

Preface

NOTE ON ORGANIZATION OF THIS BOOK

The material in this book naturally divides into two parts:

1. Part I: Chapters 1-7 cover fundamental biomechanics laws, including fluid, cardiovascular, and solid topics (1/2 semester).

2. Part II: Chapters 8-15 cover bioelectricity concepts, including circuit analysis, cell potentials, and Fourier topics (1/2 semester).

A Major Project accompanies the book to provide laboratory experience. It also can be divided into two parts, each corresponding to the respective two parts of the book. For a full-semester course, both parts of the book are covered and both parts of the Major Project are combined.

The chapters in this book are support material for an introductory class in biomedical engineering[1]. They are intended to cover basic biomechanical and bioelectrical concepts in the field of bioengineering. Coverage of other areas in bioengineering, such as biochemistry, biomaterials and genetics, is left to a companion course. The chapters in this book are organized around several fundamental laws and principles underlying the biomechanical and bioelectrical foundations of bioengineering. Each chapter generally begins with a motivational introduction, and then the relevant principle or law is described followed by some examples of its use. Each chapter takes about one week to cover in a semester-long course; homework is normally given in weekly assignments coordinated with the lectures.

The level of this material is aimed at first-semester university students with good high-school preparation in math, physics and chemistry, but with little coursework experience beyond high school. Therefore, the depth of explanation and sophistication of the mathematics in these chapters is, of necessity, limited to that appropriate for entering freshman. Calculus is not required (though it is a class often taken concurrently); where needed, finite-difference forms of the time- and space-varying functions are used. Deeper and broader coverage is expected to be given in later classes dealing with many of the same topics.

Matlab is used as a computational aid in some of the examples in this book. Where used, it is assumed that the student has had some introduction to Matlab either from another source or from a couple of lectures in this class. In the first half of the cardiovascular Major Project discussed below,

[1]At the University of Utah, this course is entitled Bioen 1101, Fundamentals of Bioengineering I.

Matlab is used extensively; therefore, the specific Matlab commands needed for this Major Project must be covered in class or in the lab if this particular part of the project is implemented.

A Major Project accompanies these chapters at the end of the book. The purpose of the Major Project, a semester-long comprehensive lab project, is to tie the various laws and principles together and to illustrate their application to a real-world bioengineering/physiology situation. The Major Project models the human systemic cardiovascular system. The first part of the problem takes approximately one-half of a semester to complete; it uses Matlab for computer modeling the flow and pressure waveforms around the systemic circulation. Finite-difference forms of the flow/pressure relationships for a lumped-element model are combined with conservation of flow equations, which are then iterated over successive cardiac cycles. The second half of the problem engages a physical electrical circuit to analyze the same lumped-element model and exploits the duality of fluid/electrical quantities to obtain similar waveforms to the first part. This Major Project covers about 80% of the topics from the chapter lectures; the lectures are given "just-in-time" before the usage of the concepts in the Major Project.

Although the Major Project included with this book deals with the cardiovascular system, other Major Project topics may be conceived and substituted instead. Examples include modeling human respiratory mechanics, the auditory system, human gait or balance, or action potentials in nerve cells. These projects could be either full- or half-semester assignments.

ACKNOWLEDGEMENTS

The overarching organizational framework of these chapters around fundamental laws and principles was conceived and encouraged by Richard Rabbitt of the Bioengineering Department at the University of Utah. Dr. Rabbitt also provided much of the background material and organization of Chapters 2 and 4. Angela Yamauchi provided the organization and concepts for Chapter 3. David Warren contributed to the initial organization of Chapter 8. Their input and help was vital to the completion of this book.

Douglas A. Christensen
University of Utah
March 2009

CHAPTER 8

Ohm's Law: Current, Voltage and Resistance

8.1 INTRODUCTION

Biological organisms rely upon myriads of electrical activities for proper functioning. These electrical processes occur continuously and are vital to life. For example, ion (charged particle) movement is responsible for all signal transmission along nerves and for all muscle contraction. Right now, as you read these words, several billions of ions are being rapidly transported back and forth across the cell membranes of the neurons in your retina, optic nerve, and brain, not to mention your heart muscle, kidney, blood vessels and all other cellular tissue. Ions commonly involved in biological activity include calcium (Ca^{++}), potassium (K^+), sodium (Na^+), chloride (Cl^-) and bicarbonate (HCO_3^-). These ions flow through the fluid environment both inside and outside the cells, as shown in stylized form in Fig. 8.1.

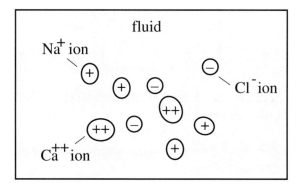

Figure 8.1: Some common ions found in the fluid inside and outside biological cells.

A major component[1] of the force causing these ions to move (or not move) across the cell wall is electrostatic, that is, the force produced on a charged particle by the presence of other nearby charged particles, some having the same electrical sign, some with opposite sign. The charge of a single electron is $q = -1.60 \times 10^{-19}$ C, where C is the symbol for coulomb, the SI unit of charge. The sign of the electron is negative, thanks in large part to choice by Benjamin Franklin. The charge of an ion is determined by how many electrons are *missing* from its atomic shell (a deficit of

---

[1]This is not the only component of force. Another important force on the ions is a diffusional force produced by concentration differences. The balancing of the electro-diffusional forces is covered in Chapter 14.

electrons, giving a positive sign to the net charge) or are *added* to the atomic shell (an excess, giving a negative sign to the net charge). For example, since a sodium ion (Na^+) is missing one electron, its charge is $+1.60 \times 10^{-19}$ C. The charge of a chloride ion (Cl^-), which has one excess electron, is -1.60×10^{-19} C, and the charge of a calcium ion (Ca^{++}), which is missing two electrons, is $+3.20 \times 10^{-19}$ C.

Bare, mobile electrons also occur in nature (although not to any great extent in biological tissues). Mobile electrons are present in large concentrations only in metals and semiconductors. In these materials, they flow as electron currents and are an essential part of modern electronic devices and circuits. In general, an **electric circuit** is composed of a source and one or more elements connected together by metallic wires and leads (usually copper, but sometimes gold, silver or aluminum) to form a closed path. Figure 8.2 is an example of a simple circuit. The Major Project that models the human systemic circulation uses an analog electric circuit in its second half.

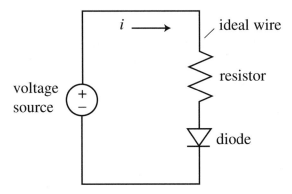

Figure 8.2: Example of an electric circuit consisting of a source, resistor and diode connected in a closed path by wires (assumed ideal, with zero resistance).

To quantitatively describe ion (and electron) movement, we first need to define the concepts of charge, electric field, current and voltage.

8.1.1   CHARGE

Figure 8.3 shows two charged particles, q_1 and q_2 (both positive in this example) in close proximity to each other. It is well known that like charges repel and unlike charges attract. This phenomenon can be used to yield a definition of both the sign and the value for any charge. Consider q_1 to be a positive source charge, and q_2 to be a test charge. The sign of q_2 is defined as positive if the force F that it experiences due to the presence of q_1 is directed away from q_1 (since q_1 is positive); it is negative if the force is directed toward q_1. The value of the charge q_2 is determined by the magnitude of the force F. The larger the force, the larger the value of the charge q_2. Coulomb's Law is an equation relating these various quantities, and will be covered in more detail in Chapter 11. As mentioned earlier, the SI unit of charge is the coulomb, with symbol C.

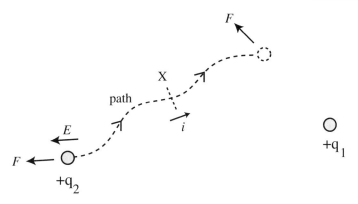

Figure 8.3: Two charges q_1 and q_2 of the same sign repel. Thus, each charge experiences a force F (which can also be represented by an electric field E). As the charge q_2 moves, it produces a current i and it may also change its potential energy, or voltage.

8.1.2 ELECTRIC FIELD

The existence of the force F on charge q_2 due to charge q_1 can be represented by an electric field E at the position of q_2, whose direction is the same as the force F (thus E is a vector, with direction and magnitude) and whose magnitude is given by $E = F/q_2$. Because of the division by q_2, the magnitude of E is not dependent on the size of the test charge q_2, but only on the size and sign of the charge q_1 (and also inversely upon the distance between the charges, as will be evident when we cover Coulomb's Law in Chapter 11). The electric field is a conservative field, as defined in Section 7.4.2 in Part I.

8.1.3 CURRENT

Now suppose that the charge q_2 in Fig. 8.3 moves along the path shown by the arrows from the lower left to the upper right. The movement of charge produces a current (denoted by the symbol i). Current is the measure of the rate of flow of charge, and it has the SI unit of ampere (A). If an amount of charge Δq passes a certain point such as X during time Δt, the current i is given by

$$i = \Delta q/\Delta t. \tag{8.1}$$

In the limit as Δt approaches zero, (8.1) becomes the differential equation

$$i = dq/dt. \tag{8.2}$$

By convention, electrons flow in the *opposite* direction to the flow of positive current, since the electron has a negative charge. This may be initially confusing, but it will become more comfortable with experience. For ions, positive current flows in the same direction as the flow of positively charged ions, but it flows in the opposite direction of negatively charged ions.

8.1.4   VOLTAGE

As the charge q_2 moves, it may move closer to q_1 (or further away). If so, its movement is against (or with) the force F. As we have seen in the previous chapter, work—a change in energy—is equal to the distance moved multiplied by the force experienced. Thus, the particle q_2 will increase its energy if it moves closer to q_1 against the force, or decrease its energy if it moves further away with the force. The former case is shown in the example of Fig. 8.3. This form of energy is properly called **potential energy**, since if q_2 is released it will accelerate and fly away from q_1, changing its potential energy into kinetic energy.

If we divide the potential energy of the charged particle by its charge, we obtain the voltage[2] v. Thus,

$$v = E_{\mathrm{p}}/q_2 \tag{8.3}$$

where E_{p} is the charge's potential energy at any given position. Voltage is the amount of work needed to move a unit charge between two points. A large voltage represents the potential to do a large amount of work[3]. The SI unit of voltage is the volt, with symbol V. From (8.3) it can be seen that the unit of volt (V) is equivalent to a joule per coulomb (J/C).

Figure 8.4 plots the change in voltage for q_2 as it moves from one position to another (in this case, closer to q_1, increasing its voltage). Since voltage is measured between two points, its value is *relative*; often in circuits, one point on the circuit is chosen to be at zero voltage, and the voltages at all other points are measured in reference to this point.

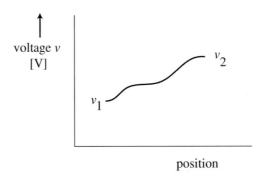

Figure 8.4: The voltage (proportional to potential energy) of the charge q_2 in Fig. 8.3 as it moves along the path. Its voltage increases because it moves closer to q_1.

---

[2]Voltage is sometimes referred to as "potential," or "electrical potential," which is understandable since it is related to potential energy.

[3]But remember, voltage is work *per unit charge*. So a large amount of charge must flow (i.e., a large current) in order to achieve this amount of work. As we will see later, power equals voltage multiplied by current.

8.2   OHM'S LAW

Figure 8.5 shows a particularly simple electrical circuit consisting of a voltage source connected across a single resistor with ideal wires. To a very good approximation, the current i flowing through the resistor is related to the voltage drop v across the resistor in a linear manner:

$$\boxed{v = i\,R} \qquad \text{Ohm's law} \qquad (8.4)$$

where the proportionality constant is the **resistance** R of the resistor. This linear relationship between current and voltage is known as **Ohm's Law**, named after German physicist Georg Simon Ohm. It holds for resistors in circuits, for resistive conductors and wires, and for the flow of electrolytes through solution, and even is used to help describe the electrical behavior of the cell wall (where the current is carried by ions)[4]. Any device or material that obeys Ohm's Law is termed "ohmic."

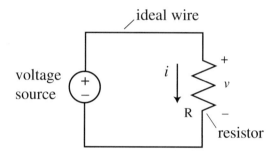

Figure 8.5: Simple electrical circuit demonstrating Ohm's Law for the resistor R.

When a plot is made of the voltage across an ohmic element as a function of the current through it, the linear relationship is obvious; see Fig. 8.6. Note that the line passes through the origin; that is, when the current reverses sign, the voltage also reverses sign. Also, if the current is zero, the voltage is zero. The slope of the line is constant and has a value equal to the resistance R of the element. R is always a positive number for passive elements like resistors. R has the SI unit of ohms, with the Greek symbol Ω. From (8.4) it can be seen that an Ω is equivalent to V/A. In some instances (for example, an ideal wire), the resistance is assumed to have a value of zero (a short circuit). On the other hand, when there is no conductive path at all between two points, the value of the resistance is infinite (an open circuit) and no current can flow.

Occasionally, the reciprocal of resistance, called the **conductance** G, is used instead of resistance. The SI unit of conductance is the siemen, or S. Conductance and resistance are related by

$$G = 1/R. \qquad (8.5)$$

[4]A useful electrical model of a cell membrane includes a capacitor in parallel with several resistors and voltage sources. The values of the resistors change dramatically as ion channels open and close, causing voltage spikes (action potentials) to propagate down nerves. This model is covered in Chapter 14.

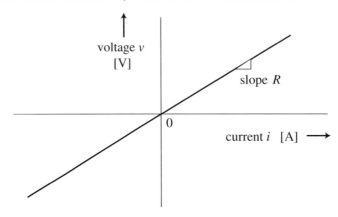

Figure 8.6: Voltage-current relationship for an ohmic element.

8.2.1   FLUID ANALOGIES

By now, it should be obvious that there is a close analogy between fluid flow (where the flowing particles are molecules of fluid) and electrical current (where the flowing particles are electrons or ions). In fact, the equations below describing the linear relationship between the driving force and the flow rate are of exactly the same form for both fluids and electrical current.

- For fluid flow though a porous membrane (Chapter 2, Part I), the driving force is the pressure difference ΔP across the membrane, and the flow Q is the fluid passing through the membrane. Darcy's Law,

$$\Delta P = Q\ R, \tag{2.5}$$

  states that the pressure is proportional to the flow, with the proportionality constant called the hydraulic or fluid resistance R, given by (2.5).

- For fluid flow through a tube (Chapter 3, Part I), the driving force is the pressure difference ΔP between the two ends of the tube, and the flow is fluid flow Q through the tube. Poiseuille's Law,

$$\Delta P = QR, \tag{3.7}$$

  shows that with laminar flow conditions, the pressure is again proportional to flow; the proportionality constant is similarly called hydraulic or fluid resistance R, given by (3.7).

- For electrical current (this chapter), Ohm's Law,

$$v = i\ R, \tag{8.4}$$

states that the voltage drop across a resistive element is proportional to the electrical current, with the proportionality constant given by the resistance R.

Note the similar form of all three equations above. Of course the quantities are different, but their mathematical behavior is identical. Conservation laws also apply to both the conservation of fluid molecules in a fluid network and the conservation of electrical charge in an electrical circuit. This is the basis for using an electrical circuit as an analogy for a fluid circuit. In particular, in the second half of the Major Project that models the human systemic circulatory system, an electrical circuit is used to model the pressures and the blood flow.

8.3   SIGN CONVENTIONS FOR VOLTAGE AND CURRENT

When doing an analysis of the resistors in circuits or resistive elements in biological models, it is important to keep accurate track of the signs of both the current and the voltage across the resistor. Figure 8.7 shows the symbol for a resistor used in schematic diagrams. A current i is assumed to flow through the resistor. The reference direction for *positive* current is the same as the direction of the arrow, shown to the side of the symbol. The voltage v across the resistor is given by the difference between its values on either side of the resistor; thus $v = v_1 - v_2$ in Fig. 8.7. (More correctly the symbol Δv should be used instead of v, but this becomes cumbersome after a while, so is usually shortened to just v.)

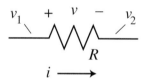

Figure 8.7: Circuit symbol for a resistor of value R. The order of the signs of v and the direction of i is given by the passive sign convention.

The order of the signs in Fig. 8.7 follows the **passive sign convention (psc)** which states that *the positive sign for the voltage reference is put on the side of the resistor where the positive current enters.* We will always follow the passive sign convention in this book.

Although the arrow points from left to right in Fig. 8.7, that does *not* mean that the current i always flows in that direction. It merely sets the reference direction of flow for *positive* current; that is, when the sign of the current i is positive, it flows in the direction of the arrow, or from left to right in Fig. 8.7 (and—hold on—electrons flow from right to left). But when the current i has a negative value, it flows in the opposite direction of the arrow, or from right to left in Fig. 8.7 (and—you guessed it—electrons flow from left to right).

Similarly, the voltage v can have either a positive or negative value. If v has a positive value, the voltage on the left side of the resistor in Fig. 8.7 is greater than the voltage on the right ($v_1 > v_2$). On the other hand, if v has a negative value, the voltage on the left is less than the voltage on the right ($v_1 < v_2$). But the signs of v and i are always linked together. From Ohm's Law[5] and the passive sign convention, when i is positive, v is also positive, and vice versa.

The direction initially chosen for the current arrow is usually arbitrary when setting up a circuit analysis. In every case, however, the passive sign convention—which relates the order of the signs of v to the direction of i—must be used in setting up the circuit references. For example, Fig. 8.8 shows a case where the positive current direction has arbitrarily been chosen to be from right to left. Note that the order of the signs for v in Fig. 8.8 has been reversed from Fig. 8.7 in order to follow the passive sign convention. Although the direction of the positive current arrow is often arbitrary, once that direction has been chosen, the order of the voltage signs must follow the passive sign convention!

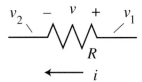

Figure 8.8: The direction of the positive current flow through the resistor R has been reversed compared to Fig. 8.7. The order of the signs of v has also been reversed to follow the passive sign convention.

8.3.1 RESISTIVITY OF BULK MATERIALS

In the previous discussions, the resistive device was considered as a whole, possessing a net resistance R. In circumstances where microscopic behavior is of more interest, it is useful to consider the resistive properties of a small incremental volume of the material making up the larger device. This property is called the **resistivity** ρ of the material, and it is specific to the material being considered, not to the particular geometry or size of the device. The SI unit of resistivity is ohm-meter ($\Omega\cdot$m). This quantity is especially useful when analyzing bulk materials such as ionic solutions or insulating layers. An insulator such as quartz has a resistivity of about $\rho = 1.0 \times 10^{+17}$ $\Omega\cdot$m. Normal saline solution has a resistivity of about $\rho = 0.20$ $\Omega\cdot$m. Metals have much lower resistivity; for copper, $\rho = 1.7 \times 10^{-8}$ $\Omega\cdot$m.

For elements with certain common geometries, there exists a simple relationship between the bulk resistivity ρ (a material property) and the overall resistance R of the element. For example, if the cylindrical resistor shown in Fig. 8.9 is composed of a homogeneous material with resistivity ρ, the net resistance of the resistor is given by[6]

$$R = \rho l / A. \tag{8.6}$$

---

[5]Remember that R is always a positive number for resistors.
[6]To practice consistency checking, use a units check and a ranging check on (8.6) to confirm that it has the correct form.

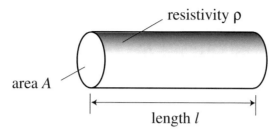

Figure 8.9: A cylinder made from material with a bulk resistivity of ρ.

Sometimes the reciprocal of resistivity, called the **conductivity** σ, is used. The two quantities are related by

$$\sigma = 1/\rho. \tag{8.7}$$

The SI units of conductivity are siemen per meter, or S/m.

8.4   DIODES AND OTHER NON-OHMIC CIRCUIT ELEMENTS

Although many electrical elements follow Ohm's Law at least over some range of their operation, other elements do not. Some circuit elements, such as the **diode**, are purposely made to behave in a non-ohmic manner.

The schematic symbol for a diode is shown in Fig. 8.10. Note that the voltage signs and current direction are set up to follow the passive sign convention.

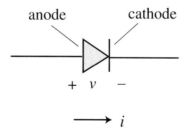

Figure 8.10: The symbol for a diode. It allows easy current flow in the direction of the arrow, but blocks flow in the opposite direction.

A diode is used as a one-way "valve" of current. It allows easy flow in the direction of the arrow (the "forward" direction, from anode to cathode) when the voltage v is positive (i.e., when the voltage on the left side in Fig. 8.10, or anode side, is more positive than on the right side, or cathode side). But the diode is highly resistant to flow in the opposite direction (the "reverse" direction) when the voltage v is negative (i.e., when the right side has greater voltage than the left side in Fig. 8.10). This asymmetric current-voltage behavior is plotted in Fig. 8.11(a) for a typical diode. Note that the current is large and positive when v is positive, equivalent to a low forward resistance, but that the current is small and negative when v is negative, equivalent to a high reverse resistance.

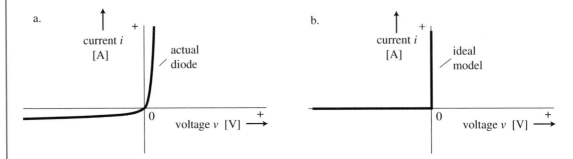

Figure 8.11: (a) The current-voltage relationship for a typical diode; (b) the idealized model of a diode, consisting of a short circuit in the forward direction and an open circuit in the reverse direction.

Often an *ideal* model of a diode is employed in circuit analysis. This makes the analysis considerably simpler. In the ideal model, it is assumed that in the forward direction, the diode possesses *zero* resistance (a short circuit); this means that there is no voltage drop across the diode regardless of the amount of current through it. However, in the reverse direction, the diode acts like an open circuit with *infinite* resistance, and no current flows regardless of the amount of voltage. This ideal model behavior is shown in the current-voltage plot of Fig. 8.11(b).

Several other important circuit elements are also non-ohmic. The transistor and the operational amplifier are examples of active devices that have gain and do not follow Ohm's Law. The operational amplifier is covered in Chapter 10.

8.5   POWER LOSS IN RESISTORS

To pass a current i through a resistor of value R requires an electrical force, measured by the voltage drop v across the resistor. Analogous to the formula for power loss in fluids [see Equation (3.9)], the electrical power loss in a resistor is given by the product of current and voltage:

$$P = iv, \tag{8.8}$$

in SI units of watts (W). Equivalent expressions can be obtained by using Ohm's Law (8.4) in (8.8):

$$P = i^2 R = v^2/R. \tag{8.9}$$

8.6   PROBLEMS

8.1. A tungsten resistance wire is 100 m long with a round cross section and a diameter of 0.20 mm. When a variable voltage is applied between its ends, a current is measured as plotted below.

a. What is the resistance R of this wire?

$$[ans: R = 180 \; \Omega]$$

b. What is the resistivity ρ of this wire?

$$[ans: \rho = 5.7 \times 10^{-8} \Omega \cdot m]$$

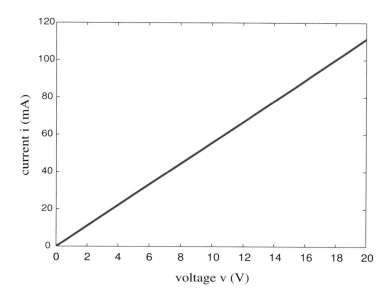

Figure 8.12: Plot of current-voltage relastionship for wire analyzed in Problem 8.1.

CHAPTER 9

Kirchhoff's Voltage and Current Laws: Circuit Analysis

9.1 INTRODUCTION

As we have seen, fluid networks consist mainly of tubes, pipes and valves connected together and to pressure sources. The tubes have various degrees of resistance, and sometimes the tubes have compliance. An excellent example of a fluid network in biology is the cardiovascular system. Similarly, electrical networks—electrical circuits—are also composed of elements connected together. In this case, the basic elements are resistors, capacitors (discussed later), inductors (also discussed later), and voltage or current sources. We will start with the three basic elements whose symbols are shown in Fig. 9.1.

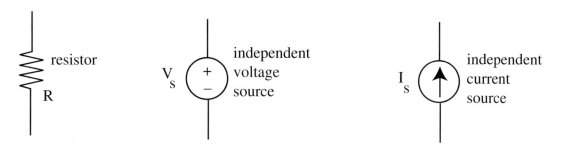

Figure 9.1: The symbols for three basic electrical elements.

The **resistor's** symbol and its ohmic voltage/current behavior have already been introduced in Ohm's Law unit. The **independent voltage source** is a source whose voltage V_s across its terminals is always the same regardless of the current through the source (it is therefore called a "constant-voltage" source). This is an ideal model, never really found in practice, but batteries and voltage power supplies come close to this behavior for low to moderate currents. The **independent current source** is a mirror image of the voltage source: the current I_s out of this source is always the same regardless of the voltage across its terminals (a "constant-current" source). Again this is an ideal model and is only an approximation of real current sources.

For every fluid network model, there is an equivalent electrical network model. For example, two very simple fluid networks are shown in Fig. 9.2 along with their electrical equivalents. The wide tubes are assumed to have negligible resistance, as are the wires.

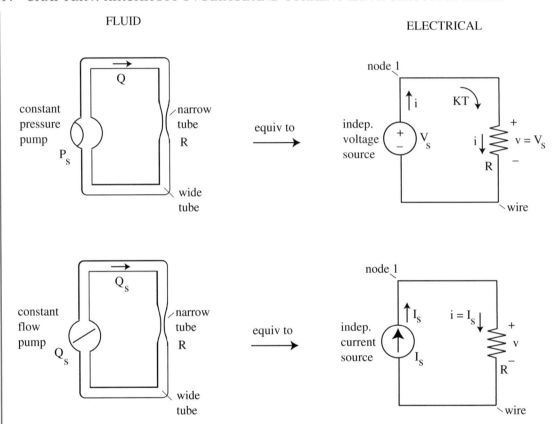

Figure 9.2: Two simple fluid networks on the left with their equivalent electrical circuits on the right.

The fluid network in the upper part of Fig. 9.2 is driven by a constant-pressure pump, such as an air-diaphragm pump; for this type of pump, P_s is constant but Q is not. Thus, if the resistance R of the narrow-tube section increases somehow, the flow rate Q will go down. Its electrical equivalent on the right, using an independent voltage source, shows the same behavior. The voltage V_s is always constant. The voltage is the same across both the source and the resistor, and the same current i goes through both the source and the resistor. By Ohm's Law, if the resistance R increases, the current i will decrease.

The fluid network in the lower part of Fig. 9.2 is driven by a constant-flow pump, such as a roller pump. The flow rate Q_s out of this source is constant, but the pressure across it is not. (Although more complex than constant-pressure pumps, a constant-flow pump is often used in hydraulic elevators, where the rate of ascent and descent needs to be constant regardless of the load inside the elevator.) If the resistance of the narrow-tube section increases, the pressure drop across the tube (and correspondingly the pump) goes up. The electrical equivalent, which uses an independent current source, has this same behavior. The voltage v is the same across both the source and the resistor, and the same current goes through both the source and the resistor. The current I_s is always constant. By Ohm's Law, if the resistance R increases, the voltage v will increase.

9.2   KIRCHHOFF'S VOLTAGE LAW (KVL)

Kirchhoff's Voltage Law applies to all electrical circuits. It states:

> *The algebraic sum of voltages around any closed path (loop) is zero.*

The application of this principle to circuits is straightforward. Going around any loop while summing its voltages is called taking a "**Kirchhoff's Tour**" (KT). The tricky part is keeping track of the signs. To help with getting the signs correct, we must always use the passive sign convention (*psc*) for resistors (see Ohm's Law in Chapter 8), and then follow Rule 1:

> **Rule 1** – When using KVL around a loop, the sign of the voltage contribution across any element is the sign *first* encountered in the direction of the tour.

Let's apply KVL to a simple example. In the upper-right electrical circuit of Fig. 9.2, start the KT at the part of the circuit labeled node 1. Go clockwise around the loop summing voltage contributions from each element (only two in this case) until you get back to the starting point. The first voltage crossed is $+v$; it is positive since the $+$ sign was encountered first when going in a clockwise direction. The next voltage term is $-V_s$; it is negative because the $-$ sign was encountered first. That gets us back to node 1, the starting point. KVL says that the sum of these voltages is zero, or

$$+ v - V_s = 0. \tag{9.1}$$

Thus, $v = V_s$, as is perhaps obvious (and was already anticipated in the figure labels). We'll use KVL for more complex examples later.

9.3   KIRCHHOFF'S CURRENT LAW (KCL)

This law also applies to all electrical circuits. It states:

> *The algebraic sum of currents at any node is zero*
> *-or- The currents entering any node equal the currents leaving.*

The strict definition of a **node** is any point at which two or more circuit elements join, although in a while we will be concerned only with nodes where *more* than two elements join (called **essential nodes**). In fluids, nodes are called junctions. Note that KCL is just a statement of the conservation of mass, identical to the rule for incompressible fluid volumetric flow. In fluids, the mass is composed of fluid molecules; in the electrical case, the mass is that of charged particles.

Once again, the application of KCL is straightforward as long as the signs of each current are accounted for correctly. Let's apply KCL to the simple example shown in the lower-right part of

Fig. 9.2. At the (nonessential) node labeled node 1, the current entering is I_s. The current leaving is i. Using KCL,

$$i = I_s, \tag{9.2}$$

as is again perhaps intuitive (and already labeled as such in Fig. 9.2).

Example 9.1.   Four-Wire Node
Another example of KCL, applied to an essential node with four connections, is shown in Fig. 9.3.

Here KCL gives

$$i_1 + i_3 = i_2 + i_4,$$

or

$$i_1 + i_3 - i_2 - i_4 = 0. \tag{9.3}$$

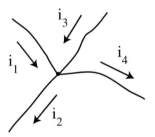

Figure 9.3: KCL applied to an essential node with four connections.

9.4   RESISTIVE CIRCUIT ANALYSIS USING THE BRANCH CURRENT METHOD

A somewhat more complex circuit is shown in Fig. 9.4. Kirchhoff's Laws and Ohm's Law can be applied to solve for the voltages and currents in all elements of this circuit by one of three possible circuit analysis methods. Here we will use the **Branch Current Method** (the other two methods will be left for later classes). The circuit of Fig. 9.4 has two essential nodes (i.e., nodes with more than two connections) labeled node 1 and node 2. A **branch** is defined as a single path that connects one essential node to another. In this circuit there are three branches connecting essential node 1 to essential node 2.

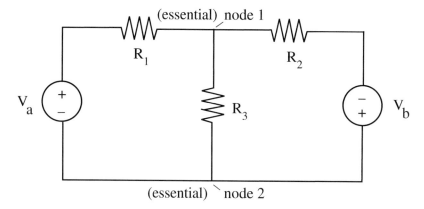

Figure 9.4: Example of a circuit to be analyzed by the Branch Current Method.

The **Branch Current Method** has four steps:

Step 1. Assign a current to each branch that does not have a current source. Give the current a name (using i symbols) and a direction. You can choose each direction arbitrarily. (Exception: for branches which contain a current source, the current i is always in the same direction and of the same value as specified by the source.)

Step 2. Assign a voltage to each element that does not have a specified voltage. Give the voltage a name (using v symbols) and arrange the signs of the voltage according to the *passive sign convention.* (Exception: sources, being active, do not have to follow the passive sign convention. The voltage value and polarity of voltage sources are always as specified by the source; the current direction through the voltage source, however, is arbitrary.)

Step 3. Apply KVL to each independent loop, Ohm's Law to each resistor, and KCL to each essential node. This results in a series of simultaneous algebraic equations.

Step 4. Solve the simultaneous equations for each desired v and i.

Let's apply these steps to Fig. 9.4. After assigning names and directions by Steps 1 and 2, we have something that looks like Fig. 9.5. Note the correct application of the *psc* in Fig. 9.5.
Now apply KVL of Step 3 to Fig. 9.5. The first KT we will take is around the left loop in the clockwise direction (denoted KT1) starting at the lower left corner. Summing voltages around this loop and following Rule 1, we get

$$- V_a + v_1 + v_3 = 0. \tag{9.4}$$

Next, take a KT around the right loop starting at node 2 and going in the clockwise direction denoted KT2. Summing voltages, we get

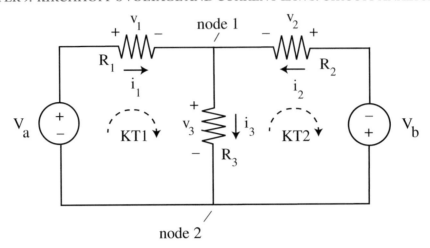

Figure 9.5: The circuit of Fig. 9.4 after current and voltage labels have been added.

$$- v_3 - v_2 - V_b = 0. \tag{9.5}$$

We'll stop KVL here. Why not apply KVL to the outer loop? Because no new equation would result that is independent of the two we already have. Thus, Rule 2:

Rule 2 – Each independent loop must include at least one *new* branch not included in any other loop.

Now apply Ohm's Law to each resistor:

$$v_1 = i_1 R_1,$$
$$v_2 = i_2 R_2,$$
$$v_3 = i_3 R_3. \tag{9.6}$$

Note that the *psc* has automatically taken care of signs. But before we go any farther, we can immediately put the three Equations (9.6) into their respective places in (9.4) and (9.5) to get

$$+i_1 R_1 + i_3 R_3 - V_a = 0. \tag{9.7}$$
$$-i_2 R_2 - V_b - i_3 R_3 = 0. \tag{9.8}$$

In fact, in all future KT's we'll be able to write down equations like (9.7) and (9.8) *directly* using Ohm's Law and KVL together, thus skipping the intermediate step (9.6).

Now use KCL at node 1 of Fig. 9.5 to get

$$i_1 + i_2 = i_3. \tag{9.9}$$

We'll stop KCL here. Why not apply KCL to the other node? Because no new equation would result that is independent of the one we already have. Thus, Rule 3:

Rule 3 – For n nodes, KCL gives $(n - 1)$ independent equations.

Now we go to Step 4. We have three independent simultaneous Equations [(9.7), (9.8), and (9.9)] with three unknowns [i_1, i_2, and i_3]. There are various ways to solve them (such as Kramer's Rule and Matlab), but we'll simply use substitution. Solving for i_1 from (9.9):

$$i_1 = i_3 - i_2. \tag{9.10}$$

Substituting this into (9.7) gives

$$- i_2 R_1 + i_3 (R_1 + R_3) = V_a, \tag{9.11}$$
$$- i_2 R_2 - i_3 R_3 = V_b. \tag{9.12}$$

Dividing (9.11) by R_1, and (9.12) by $-R_2$, gives

$$- i_2 + i_3 ((R_1 + R_3)/R_1) = V_a/R_1, \tag{9.13}$$
$$+ i_2 + i_3 (R_3/R_2) = -V_b/R_2. \tag{9.14}$$

Adding (9.13) to (9.14) so that i_2 cancels, then solving for i_3 gives

$$i_3 = \dfrac{\dfrac{V_a}{R_1} - \dfrac{V_b}{R_2}}{\dfrac{R_1 + R_3}{R_1} + \dfrac{R_3}{R_2}}. \tag{9.15}$$

Now that we have an equation for i_3, we could find the equations for i_1 and i_2 if desired. We could also find equations for v_1, v_2, and v_3 easily by multiplying the currents by their respective resistances. Also if values have been given for the various components, we could calculate the values of all currents and voltages in the circuit. An example using another circuit is given next.

Example 9.2. Another Circuit Analysis Using Kirchhoff's Laws
In the circuit below,
a. Solve for i_2 in terms of circuit components.
b. Evaluate i_2 and v_2 for the component values given.

$$I_s = 1 \text{ mA} \qquad V_a = V_b = 10 \text{ V}$$
$$R_1 = 10 \text{ k}\Omega \qquad R_2 = 10 \text{ k}\Omega \qquad R_3 = 20 \text{ k}\Omega$$

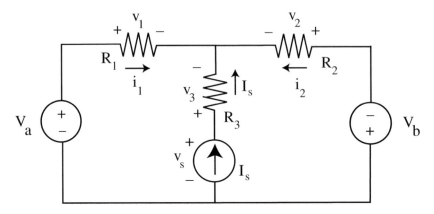

Figure 9.6: Circuit analyzed in Example 9.2.

Solution
a. Steps 1 and 2 of the Branch Current Method have already been done in the figure above. Note that the current through R_3 has been set to I_s, since it is forced to be that value by the current source in that branch. Now proceed with Step 3:
KVL around the left loop:

$$- V_a + i_1 R_1 - I_s R_3 + v_s = 0. \tag{9.16}$$

KVL around the right loop:

$$- v_s + I_s R_3 - i_2 R_2 - V_b = 0. \tag{9.17}$$

KCL at upper node:

$$i_1 + i_2 + I_s = 0. \tag{9.18}$$

Step 4: Now solve these equations. Add (9.16) to (9.17), immediately. Several terms cancel, leaving

$$- V_a + i_1 R_1 - i_2 R_2 - V_b = 0. \tag{9.19}$$

Solve for i_1 from (9.18):

$$i_1 = -I_s - i_2. \tag{9.20}$$

Put (9.20) into (9.19):

$$- V_a - I_s R_1 - i_2 (R_1 + R_2) - V_b = 0. \tag{9.21}$$

This can be solved directly for i_2:

$$i_2 = -(V_a + V_b + I_s R_1)/(R_1 + R_2). \tag{9.22}$$

b. Putting the values from Fig. 9.6 into (9.22) gives

$$i_2 = -(10 + 10 + 10)V/(10 + 10) \text{ k}\Omega = -3/2 \text{ mA} = -1.5 \text{ mA},$$

so

$$v_2 = i_2 R_2 = (-1.5 \text{ mA})(10 \text{ k}\Omega) = -15 \text{ V}. \tag{9.22}$$

What does the negative sign for i_2 mean? It says that the current in R_2 is really going in the opposite direction of the arrow, or left to right. We guessed wrong when we set up the current arrows, but it's fine. The signs of the answers will give the correct final directions. What does the negative sign for v_2 mean? It says that the magnitude of the voltage on the left side of R_2 is really larger than the voltage on the right side. Once again, the signs of the answers will give the correct polarity.

9.5   PROBLEMS

9.1. Let $R = 2.0$ kΩ in the circuit of Fig. 9.7. Mark on the figure the correct polarity for the voltage v using the passive sign convention, then use KVL and Ohm's Law to solve for the current i and the voltage v.

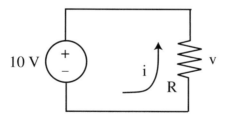

Figure 9.7: Circuit to be analyzed in Problem 9.1.

[ans: $v = -10$ V   and $i = -5.0$ mA]

9.2. a. Use the Branch Current Method (KVL and Ohm's Law) to solve for the current i and the voltage v_2 across the 1.0 kΩ resistor in the circuit of Fig. 9.8.

[ans: $i = -2.0$ mA   and $v_2 = -2.0$ V]

b. How many electrons will pass point a in the circuit during 5.0 seconds?

[ans: 6.3×10^{16} electrons]

c. Is the direction of the electron flow clockwise or counterclockwise?

[ans: clockwise]

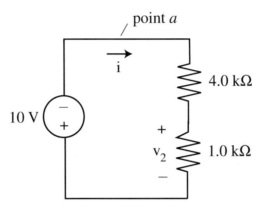

Figure 9.8: Circuit to be analyzed in Problem 9.2.

9.3. a. For the circuit shown in Fig. 9.9, use the Branch Current Method to find an expression for i_3 in terms of the circuit components.

$$[\text{ans: } i_3 = -V_a(R_1 + R_2)/(R_1 R_2 + R_2 R_3 + R_1 R_3)]$$

b. Do at least two ranging checks on the answer of part **a.**

c. Evaluate the voltage across R_3 for the component values given.

$$[\text{ans: } v = -600 \text{ mV}]$$

9.4. a. Use the Branch Current Method to derive an expression for v_3 in the circuit of Fig. 9.10 in terms of the other parameters of the circuit. (Hint: Solve for i_3 first.)

$$\left[\text{ans: } v_3 = \left[\frac{V_b R_1 - V_a (R_1 + R_2)}{R_1 R_2 + R_1 R_3 + R_2 R_3} \right] R_3 \right]$$

b. Perform a units check on this equation.

c. Perform one ranging check on this equation.

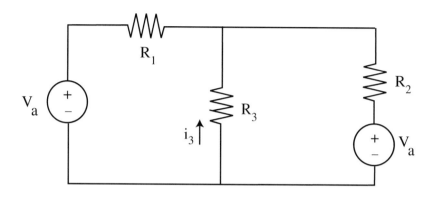

Values: $V_a = 1.0 \ V$ $R_1 = 100 \ \Omega$ $R_2 = 200 \ \Omega$ $R_3 = 100 \ \Omega$

Figure 9.9: Circuit to be analyzed in Problem 9.3.

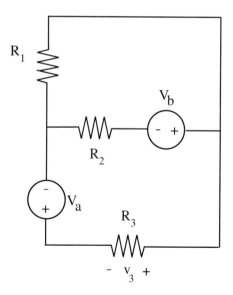

Figure 9.10: Circuit to be analyzed in Problem 9.4.

CHAPTER 10

Operational Amplifiers

10.1 INTRODUCTION

There are several sources of electrical signals produced inside the body: signals from cardiac muscle (recorded as electrocardiograms, ECG or EKG), skeletal muscle (electromyograms, EMG), and brain activity (electroencephalograms, EEG). These voltage signals can be measured with electrodes on the surface of the body, that is, *noninvasively*, to diagnose diseases and determine the physiological state of various organs. But the voltages are very weak by the time they propagate to the body's surface— usually on the order of just a few microvolts (μV). By the time they reach the measurement and recording devices, they are even smaller. Figure 10.1(a) models an ECG signal on the surface of the body as an independent voltage source V_{ecg}. This model is simplified, because in reality the ECG signal is time-varying and not constant, but at any given time it may be approximated by this model. Electrodes and wires connect the source to a measurement device (for example, an oscilloscope or an analog-to-digital converter). The resistance of the wires and electrodes is modeled by a resistor R_1 and the input resistance (also known as the input impedance) of the measurement device is modeled by a resistor R_2.

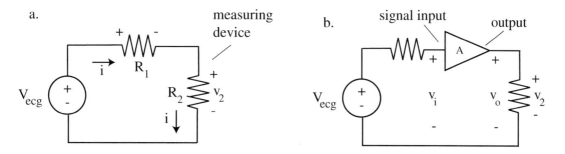

Figure 10.1: (a) Measurement of an internal electrical source V_{ecg} with electrodes on the body's surface. The wires and electrodes are modeled by R_1, and the input impedance of the measurement device by R_2. The measured voltage is v_2. (b) An amplifier with gain A (denoted by the triangle) is inserted between the source and the measurement device.

The measured voltage v_2 can be found using Kirchhoff's Laws and Ohm's Law to analyze the simple circuit of Fig. 10.1(a). KCL applied to any (nonessential) node around the circuit shows that the current i must be the same in all segments of the circuit. This has already been noted in the figure. KVL around the loop, starting at the lower-left corner and going clockwise, gives

$$-V_{ecg} + i R_1 + i R_2 = 0, \tag{10.1}$$

so

$$i = V_{ecg}/(R_1 + R_2). \tag{10.2}$$

From Ohm's Law,

$$v_2 = i R_2 = \left(\frac{R_2}{R_1 + R_2} \right) V_{ecg}. \tag{10.3}$$

Note in (10.3) that the voltage v_2 has been reduced compared to the source voltage V_{ecg} by the ratio $R_2/(R_1 + R_2)$. This is an example of a **voltage divider** circuit, which occurs whenever a source voltage is applied to a string of resistors in series, and the measured voltage is taken across one of these resistors. Since the ratio in the parenthesis of (10.3) is always less than unity (unless R_1 happens to be zero), the voltage v_2 is less than V_{ecg}. Since V_{ecg} is already small, v_2 is even smaller and can be difficult to measure in the presence of noise. Some means of amplifying the source voltage is needed.

Figure 10.1(b) shows a simple amplifier with a voltage gain of A (and a very high input impedance) inserted between the source and the measuring device. The amplifier in this example has a single input, denoted by voltage v_i. Its output voltage v_o is related to the input voltage by

$$v_o = A \, v_i. \tag{10.4}$$

The voltage gain A may be large (up to perhaps 1×10^5). Thus, the measured voltage can be boosted with the amplifier to the neighborhood of a volt or a significant fraction of a volt, which is easy to measure and record.

10.2   OPERATIONAL AMPLIFIERS

A very popular and versatile type of amplifier is the operational amplifier, or **op amp**. It forms the building block for a number of convenient amplifier circuits. The device itself consists of a tiny integrated circuit (IC) with many transistors, diodes, resistors and capacitors performing the amplifying function. We will not be concerned here with the internal electronics of the op amp. Instead we treat it as a module, and model it with an equivalent circuit that describes its overall electrical behavior.

The physical package of a typical op amp is much smaller than a postage stamp, and from the outside it looks like a "bug" with wire legs, at least in the popular dual-in-line (DIP) package, or "chip". Figure 10.2 shows how an 8-pin DIP package looks from a top view. There is always some registration mark (a dot or a cutout) at one end to orient the numbering order of the pins. By convention, the pin numbers start from 1 on the left of the mark and increase in the counterclockwise

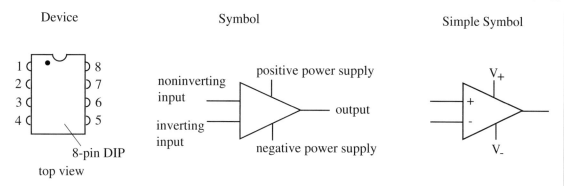

Figure 10.2: The physical layout of an example of an op amp package. Also shown are typical symbols used in op amp circuit diagrams.

direction as seen from the top, shown in Fig. 10.2. An 8-pin DIP package can actually house two separate op amps, which is common.

The complete symbol for one op amp is given at the center of Fig. 10.2. Note that the op amp has two inputs (labeled inverting and noninverting) and one output, so it is a type of dual-input, or differential, amplifier. Since the output voltage is often larger than the input voltage, the output power is often much larger than the input power. The conservation of energy principle requires that there be some other external source of power supplied to the amplifier to provide this increase in power. This takes the form of one or (usually) two power supplies that must be connected to the amplifier for it to function properly. These are labeled as the positive power supply and the negative power supply connections on the symbolic diagram.

An abbreviated symbol is often used in op amp circuit diagrams, shown on the right of Fig. 10.2. The location of the inverting input is denoted by a − sign, and the noninverting input by a + sign. It is important to note that the + and − signs inside the symbol for the two inputs have nothing to do with the actual polarity of the input voltages; they merely denote the inverting and noninverting nature of the two inputs, explained shortly. The power supply voltage leads are labeled V_+ and V_-. Here the + and − signs *do* denote the polarity of the supply voltages.

A diagram of how an op amp is hooked up is given in Fig. 10.3. The power supplies can be either bench-type voltage supplies or batteries (for portability); the positive supply requires a constant positive voltage (usually +12 V or +15 V) and the negative supply requires a constant negative voltage (usually −12 V or −15 V). These are denoted $+V_{cc}$ and $-V_{cc}$. The input voltage to the noninverting terminal is labeled v_p and the input voltage to the inverting terminal is labeled v_n. Both are referenced to the bottom wire, which is usually tied to ground at zero volts. The output voltage is v_o, again referenced to ground. The ground is indicated by a series of short lines inside an inverted triangle.

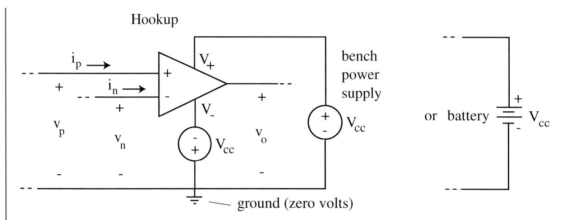

Figure 10.3: The hookup of an op amp.

We now need an equivalent circuit model to represent how the output voltage is related to the two input voltages. One version of an op amp equivalent circuit model is given in Fig. 10.4. (There are much more complex models that can be used, including such effects as offset input voltages and capacitances, but they are more detailed than we require here.) In this model, the two inputs are joined together by a large input resistance R_i. The output voltage is generated by a *dependent* voltage source in the shape of a diamond (discussed in the next section). Its voltage $A(v_p - v_n)$ is proportional to the *difference* between the noninverting input voltage v_p and the inverting input voltage v_n. The proportionality constant is the gain A. This dependent voltage source is connected to the output terminal through a small series output resistor R_o.

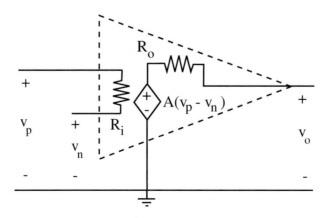

Figure 10.4: Equivalent circuit model for an op amp.

Typical values for a good op amp are listed in Table 10.1 along with simplifying approximations to these values. The input resistance R_i is usually very large—several hundreds of megohms. Thus, it can be approximated as an infinite resistance, or open circuit. The output resistance R_o is usually small—less than 75 ohms—compared to other resistors in the circuit. Thus, it can be approximated as zero resistance, a short circuit or an ideal wire. The voltage gain A is very large. This leads to a further simplification in the model, as will be apparent shortly.

Table 10.1: Op Amp Parameters	
Typical Values	Approximate Values
$R_i \approx 10^{12} \ \Omega$	$R_i \to \infty$
$R_o < 75 \ \Omega$	$R_o \to 0$
$A = 10^5 - 10^6$	$A \to \infty$

When these approximations are made in the equivalent circuit, the *ideal* op amp model of Fig. 10.5 results. In the ideal model, the inputs both terminate in open circuits in the amplifier; therefore no current can flow into these input terminals, regardless of the input voltage. Also the output is directly connected to the dependent voltage source (since $R_o = 0$), so the output voltage is the same as the dependent source voltage. From now on, we will use this ideal model in the analysis of various amplifier configurations.

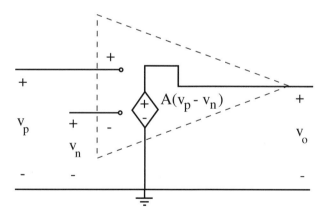

Figure 10.5: The *ideal* op amp equivalent circuit.

10.3 DEPENDENT SOURCES

We have just introduced a new type of electrical source, the dependent source. Independent sources were described in Chapter 9, where it was seen that an independent source (for example, an independent voltage source) is characterized by the fact that its output voltage is fixed at a constant value. A battery is an example of this kind of source. Independent sources are identified by the shape of a circle or by a battery symbol.

Dependent sources, on the other hand, have outputs that are variable, depending upon the value of some voltage or current in another part of the circuit. Dependent sources are identified by a diamond shape. There are four possible types, categorized in Fig. 10.6.

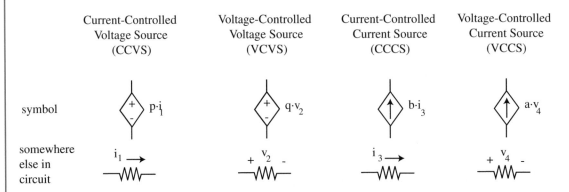

Figure 10.6: Dependent sources of four possible types.

The current-controlled voltage source (CCVS) is a dependent source whose output voltage is proportional to the current i_1 in some element in another part of the circuit. The output voltage of this source is $p \cdot i_1$ regardless of the current through the source. The proportionality constant is p, with units of V/A. The voltage-controlled voltage source (VCVS) is a source whose output voltage is proportional to the voltage v_2 across some element in another part of the circuit. The output voltage of this source is $q \cdot v_2$ regardless of the current through the source. The proportionality constant is q, with units of V/V, so it is dimensionless. The VCVS inside the ideal op amp equivalent circuit is an example of this type of source, extended such that the output voltage is proportional (with proportionality constant A) to the *difference* $(v_p - v_n)$ of two voltages on other parts of the circuit. To complete the possible dependent source configurations, Fig. 10.6 also shows a VCCS and a CCCS.

10.4   SOME STANDARD OP AMP CIRCUITS

Op amps are used in hundreds of different applications and configurations (for example, the cardiovascular Major Project utilizes one such specialized circuit: the capacitance-multiplier circuit), but there are three or four standard op amp configurations that are used again and again. They are described and analyzed next.

10.4.1 INVERTING AMPLIFIER

This arrangement is used when it is desired to amplify and invert an input signal so that the output voltage is larger than the input but of opposite polarity. Its circuit diagram is shown in Fig. 10.7(a).

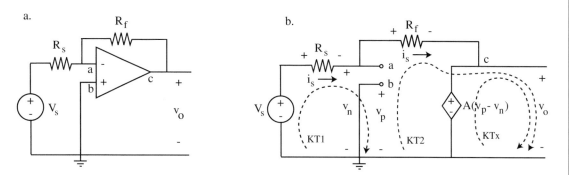

Figure 10.7: (a) Inverting amplifier; (b) to analyze, the circuit is *always* redrawn, replacing the op amp with its ideal equivalent circuit.

R_s is the resistor connecting the source to the op amp. R_f is a "feedback" resistor connecting the output back to the inverting input. The feedback resistor must always be connected back to the inverting input, *not* the noninverting input. Otherwise the amplifier will go unstable. Note in Fig. 10.7(a) that the inverting input is above the noninverting input; this order may vary from diagram to diagram.

To analyze this circuit, we *always* redraw the circuit, replacing the op amp symbol with its (ideal) equivalent circuit. We then can use the Branch Current Method or any appropriate method to find the output voltage in terms of the signal voltage. The redrawn circuit is shown in Fig. 10.7(b). v_n, v_p, and v_o are identified at their respective terminals, labeled a, b, and c.

We first use KCL at the node near a. Its result is obvious: since no current can enter terminal a, the current entering the node must equal the current leaving, or

$$i_s = i_s. \tag{10.5}$$

This result is so intuitive that from now on, whenever there is a continuous branch with multiple elements but containing *no* essential nodes (i.e., no places with a third wire branching off to the side), we will let the current be the same throughout that continuous branch.

KVL can be applied to the far-right loop—the one that includes the dependent source. This loop is indicated by the dotted line labeled KTx. Starting in the lower-left corner, we get

$$- A(v_p - v_n) + v_o = 0. \tag{10.6}$$

so

$$v_p - v_n = v_o/A. \tag{10.7}$$

Now a very important approximation can be made. Since in the ideal op amp model, $A \to \infty$, (10.7) shows that

$$v_p - v_n \approx 0, \tag{10.8}$$

or

$$\boxed{v_p = v_n} \qquad \text{Every ideal op amp} \tag{10.9}$$

Therefore, in the ideal model, the voltages at the inverting and noninverting input terminals are the same! We will employ this simplification from now on, and it makes the analysis of op amps *much* easier, as we will see. Also, when the relationship in (10.9) is used, we will never need (or want) to take a KT through the dependent source again.

Now apply KVL around the loop labeled KT1. Start at the lower left and go clockwise. The first voltage term encountered is the signal voltage V_s. The next term is the voltage across the resistor R_s, which is $+i_s R_s$. Now we need to get from terminal a (with voltage v_n) to terminal b (with voltage v_p). Since (10.9) states that $v_p = v_n$, there is *no* voltage drop encountered when going from point a to point b. So there is a *zero* contribution to the voltage sum here. That takes us to point b. There is a wire from this point back to the start of the loop (note that this means that $v_p = 0$ in this circuit), so the total KT1 tour results in:

$$- V_s + i_s R_s + 0 = 0, \tag{10.10}$$

or

$$i_s = V_s/R_s. \tag{10.11}$$

There is one branch we have not yet used in KVL, the one containing R_f, so we apply KVL to the loop labeled KT2. Note that KT2 avoids going through the dependent source to make things simpler, and instead goes through v_o. Applying KVL to KT2 gives

$$+ i_s R_f + v_o = 0. \tag{10.12}$$

Putting (10.11) in (10.12) and solving for v_o gives

$$\boxed{v_o = - \left(\frac{R_f}{R_s} \right) V_s.} \qquad \text{Inverting amp} \tag{10.13}$$

We check the units of (10.13) for consistency, and since both sides have the units of volts, we know that the derivation hasn't gone terribly astray somewhere.

Equation (10.13) states that the output voltage of this amplifier is equal to the input voltage multiplied by the ratio of R_f to R_s. This ratio gives the absolute value of the gain of this circuit, and is set by resistance values (and is therefore more stable compared to other electronic ways of setting the gain). The gain can be set to be low, moderate, or very high, as desired, by choosing the proper resistors, but cannot be larger than the native (open-loop) gain A of the op amp. The negative sign means that the polarity of the output is inverted from that of the input (thus the name "inverting" amplifier); for example, if $R_f/R_s = 100$ and $V_s = +6$ mV, the output voltage is $v_o = -600$ mV.

10.4.2 NONINVERTING AMPLIFIER

This configuration is used when it is desired that the output voltage have the same polarity as the input voltage. The original circuit is shown in Fig. 10.8(a). The circuit redrawn with the ideal equivalent circuit substituted for the op amp symbol is shown in Fig. 10.8(b). This configuration is similar to the inverting amplifier, but the signal voltage source is now connected to the noninverting input rather than the inverting input.

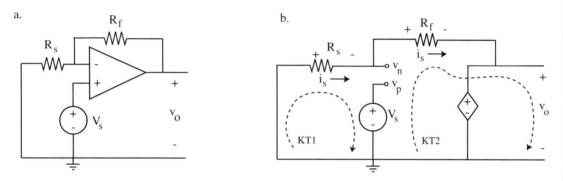

Figure 10.8: (a) Noninverting amplifier; (b) redrawn circuit using ideal op amp equivalent circuit.

We don't need to use KCL explicitly here, since there are no branching currents. KVL around KT1 gives

$$+ i_s R_s + V_s = 0, \tag{10.14}$$

so

$$i_s = -V_s/R_s. \tag{10.15}$$

In getting (10.14) we have already used the fact that $v_p = v_n$. Next, we do not need or want to take a loop through the dependent source. What's left is KVL around KT2, giving

$$- V_s + i_s R_f + v_o = 0. \tag{10.16}$$

Solving for v_o and using (10.15) results in

$$\boxed{v_o = V_s \left(1 + \frac{R_f}{R_s}\right).} \qquad \text{Noninverting amp} \tag{10.17}$$

This result shows that the gain of the noninverting amplifier is also related to the ratio of resistances, as was the case with the inverting amplifier, but with a slightly different form due to the first term in the parentheses. The gain of the noninverting amplifier is always greater than unity, even for small ratios of R_f/R_s. When $R_f >> R_s$, the gains of the two configurations are essential the same in absolute magnitude. Note that the sign of the gain for this noninverting amplifier is positive, so the input and the output voltages have the same sign (as the name "noninverting" indicates). For example, if $R_f/R_s = 10$ and $V_s = +6$ mV, the output voltage is $v_o = +66$ mV.

10.4.3   VOLTAGE FOLLOWER

This op amp configuration is the simplest possible one, and it is often used when it is necessary to avoid "loading down" a source. The term "loading" refers to the current that is required from a source when it is connected to a measuring or recording device, called the load. If the resistance of the load is low, then an appreciable amount of current will be drawn from the source. If the resistance of the source is not small (it often isn't, especially in the case of chemical electrodes), then this current draw will lower the effective voltage that can be measured at the load. A demonstration of this loading effect was seen in the example of Fig. 10.1 (the voltage-divider phenomenon). For instance, if the internal resistance of a chemical electrode is $R_1 = 1$ MΩ and the input resistance of the measuring device is $R_2 = 1$ kΩ, then (10.3) shows that only about 1/1000 of the electrode voltage will actually be seen at the terminals of the measurement device. In order for the measured voltage to be nearly as large as the electrode voltage, R_2 must be $>> R_1$.

The voltage-follower configuration accomplishes this inequality while at the same time preserving the signal voltage in both sign and magnitude. In other words, the voltage follower has a very high input impedance and a voltage gain of positive unity. Figure 10.9(a) shows the op amp circuit and Fig. 10.9(b) is the redrawn circuit with the ideal equivalent circuit.

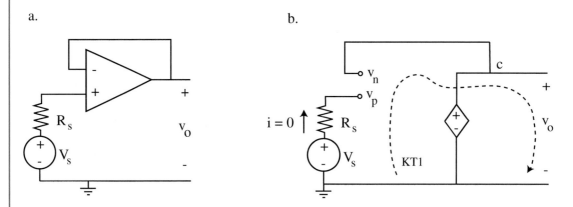

Figure 10.9: (a) Voltage follower circuit; (b) the circuit redrawn using the ideal equivalent circuit.

The key to analyzing this circuit is to note that since the branch that contains the signal source terminates in an open circuit at the op amp input, *no* current comes from the source. That is, $i = 0$ in the source branch. In turn, this means that there is no voltage drop (from Ohm's Law) across the source resistance R_s (in fact, R_s has no effect at all on the output of this circuit). Then using KVL around KT1 gives

$$-V_s + v_o = 0, \tag{10.18}$$

or

$$\boxed{v_o = V_s.} \qquad \text{Voltage follower} \tag{10.19}$$

This result confirms that the voltage follower's output voltage is exactly the same as the signal voltage, thus the name "voltage follower." This circuit is also sometimes called a unity gain amplifier. In addition, due to the (approximately) infinite input impedance of the op amp, no current is drawn from the source and there are no loading effects. Therefore, this circuit is also often called a buffer amplifier.

A slightly more complex op amp configuration is treated in the next example.

Example 10.1.   The Summing Amplifier

This amplifier is used in situations where the voltages from two or more sources are to be added together, with perhaps different gains for each voltage. Its circuit is shown in Fig. 10.10.

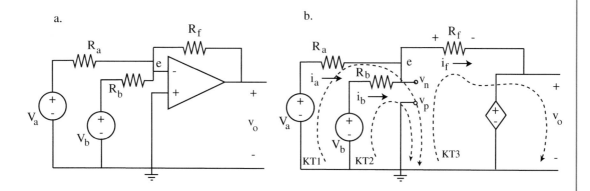

Figure 10.10: Summing Amplifier analyzed in Example 10.1.

Solution

Since we have an essential node at the node labeled e, KCL must be applied at this node to give

$$+i_a + i_b - i_f = 0. \tag{10.20}$$

Now apply KVL around KT1:

$$-V_a + i_a R_a = 0. \tag{10.21}$$

Apply KVL around KT2:

$$-V_b + i_b R_b = 0. \tag{10.22}$$

Solving for i_a and i_b from (10.21) and (10.22), and substituting in (10.20) gives

$$i_f = (V_a/R_a) + (V_b/R_b). \tag{10.23}$$

Now use KVL around KT3, which includes the remaining branch, to get

$$+i_f R_f + v_o = 0, \tag{10.24}$$

or

$$\boxed{v_o = -i_f R_f = -V_a \left(\frac{R_f}{R_a}\right) - V_b \left(\frac{R_f}{R_b}\right).} \qquad \text{Summing amp} \tag{10.25}$$

Thus, the output voltage is the weighted sum of the input voltages (with inverted signs), where the weighting factors in the parentheses depend on the values of the resistors chosen. For example, if $R_f = R_a = R_b$, then $v_o = -(V_a + V_b)$.

10.5 PROBLEMS

10.1. Use the Branch Current Method to analyze the circuit in Fig. 10.11. Remember to redraw the circuit using the ideal op amp equivalent circuit.

    a. Based on the equivalent circuit and Ohm's Law, what is the voltage across the 330 Ω resistor?

$$[\text{ans: } v = 0]$$

    b. What is the output voltage v_o?

$$[\text{ans: } v_o = -5.4 \text{ V}]$$

10.2. Use the Branch Current Method to analyze the circuit in Fig. 10.12. (Remember to redraw the circuit.) What is the output voltage v_o?

$$[\text{ans: } v_o = -9.0 \text{ V}]$$

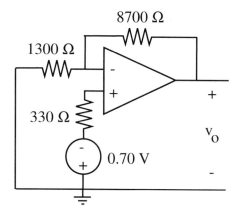

Figure 10.11: Op amp circuit to be analyzed in Problem 10.1.

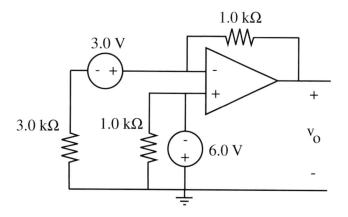

Figure 10.12: Op amp circuit to be analyzed in Problem 10.2.

CHAPTER 11

Coulomb's Law, Capacitors and the Fluid/Electrical Analogy

11.1 COULOMB'S LAW

We've seen earlier how electrical charges with the same sign repel each other and charges with opposite signs attract (this is usually stated "like charges repel and unlike charges attract"). We'll now add some detail in describing this phenomenon, and show how it leads to the concept of an electrical capacitor. Two charged particles are diagrammed in Fig. 11.1. The forces F exerted on each charge act in a direction along a line between the two charges. If F has a positive value, the forces tend to push the particles apart. If F has a negative value, the forces tend to pull the particles together.

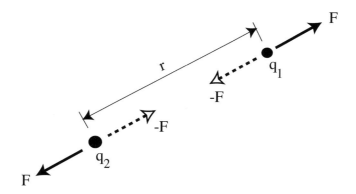

Figure 11.1: The force F between two charges depends on their charge, their separation, and the medium between them.

Coulomb's Law is a mathematical formulation of this principle:

$$F = \frac{q_1 q_2}{4\pi \varepsilon r^2},$$

(11.1)

where q_1 and q_2 are the values of the two charges (units: C) including sign,
r is the separation (m), and
ε is the permittivity of the medium between the charges.
The permittivity of free space (a vacuum) is

$$\varepsilon_0 = 8.854 \times 10^{-12} C^2/(N \cdot m^2). \tag{11.2}$$

The permittivity of other media, such as insulating materials called dielectrics, is larger than that of free space by a factor known as the relative permittivity ε_r of the material; ε_r is also called the relative dielectric constant. For example, the relative permittivity of polystyrene (for slowly varying voltages) is 2.55[1]. For a material with a relative permittivity of ε_r, then $\varepsilon = \varepsilon_0 \varepsilon_r$.

Note from Coulomb's Law that the closer the charges are to each other (the smaller r), the greater the force. This is similar to the inverse distance relationship in Newton's law of gravity. Also note that when q_1 and q_2 are of opposite sign, the force is negative and inward; when q_1 and q_2 have the same sign, the force is positive and outward. A popular demonstration of this is when a student places her hand on one electrode of a Van de Graaff generator (a generator of static electricity that is a source of excess electrons on one electrode). Because they repel each other, the excess electrons will rapidly spread out over her entire body—including her hair—causing her hair to literally "stand on end."

11.2   CAPACITORS

We are now able to address the following question: Can we store excess charges in some chosen place in a circuit in spite of Coulomb's Law, which says that they will repel each other, especially when packed densely with small r values? Let's try to store excess electrons in a metal wire. A metal is composed of atoms whose outer electrons can escape easily from the atomic shell, and therefore can move under a force to produce a current. These negatively charged electrons are called "free" or "conduction" electrons. But they leave behind the immobile positively charged shell of the atoms (as ions) that are fixed in place by the lattice of the metal. Therefore, a free electron cannot move very far away from its fixed ion (due to the attractive force described by Coulomb's Law) *unless* it is replaced by another free electron entering the vicinity of the shell to take its place. The whole metal wire is electrically neutral since the number of free electrons is equal to the number of fixed atomic cores.

Figure 11.2 shows such a metal wire, with − signs, or e^- symbols, representing free electrons and + signs representing the fixed metal ions. What would happen if we forced another (that is, an excess) electron into one end of the wire? Its negative charge is not balanced by a positive ion, so a Coulomb's repulsive force would be exerted on neighboring electrons, which would move slightly and in turn exert forces down the wire on other neighboring electrons, so on down the wire. These forces would result in one electron being expelled out the other end of the wire (assuming the wire is connected to a conductive path), and current flows.

Because there was no increase in the number of free electrons in the wire, there was no storage. Essentially, excess charges cannot be stored in normal wires.

[1]Living tissue can also be characterized by permittivity and conductivity values, usually specified for alternating (ac) electromagnetic fields in the kHz to GHz frequency range. For example, at 1 MHz, muscle has a conductivity of about 0.5 S/m and a relative permittivity of about 500.

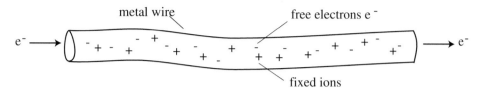

metal wire   free electrons e⁻

e⁻ ⟶   ⟶ e⁻

fixed ions

Figure 11.2: A metal wire with an excess electron forced into one end, causing expulsion of an electron from the other end (a current).

The situation is different in the arrangement of Fig. 11.3, showing a parallel-plate capacitor. In this configuration, two metal plates of area A are facing each other separated by a very thin layer of insulating material of thickness d, such as polystyrene, mylar, or mica. Wires connect to each plate, but charges cannot pass across the insulating layer. Can we store excess charges on these plates? At first you might say no, because the concentration of charges would cause very large repulsion forces, prohibiting the storage of the charges close together as in the wire example discussed above. But if there was some nearby attraction forces, these would partially cancel out the repulsion forces and storage may be possible.

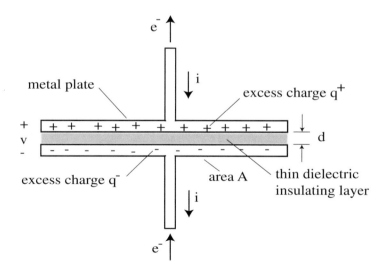

Figure 11.3: A parallel-plate capacitor (side view) that can store excess charge on one plate because the opposing plate holds the opposite charge.

Indeed, in the parallel-plate capacitor, when negative charges (electrons) are placed on one plate (the lower plate in Fig. 11.3), electrons are forced off the other (upper plate), passing through the wire connected to this plate and producing a current i. The absence of electrons on the upper plate leaves behind a concentration of positively charged fixed ions on the upper plate, exactly matching the concentration of electrons on the lower plate. Since these two concentrations are close to each other (at a distance d) and are of opposite charge, there is an attractive Coulomb's force which partially balances the repulsive force due to the concentration of electrons and helps keep the charges in place!

Now think of adding one more electron (or a charge q^-) through the lower wire and onto the lower plate. It would take some force (measured by voltage) to do so, because we would be adding to the concentration of electrons already stored there. But the voltage is not nearly as large as it would be if there were no compensating positive charges nearby. So we can postulate that the voltage v required to add another charge q might take the form

$$v = \frac{qd}{A\varepsilon}. \tag{11.3}$$

This form for the equation is justified by the following reasoning: The required voltage is proportional to the amount of charge q added, as is intuitive, and v is smaller if d is smaller since the opposing charge is closer. v is also smaller if A is larger because the charge is not being concentrated as much on the plate. The permittivity ε of the dielectric layer helps increase the attractive force across the layer, as (11.1) shows. The larger ε, the less voltage that is required, or conversely, for a given voltage, the more charge can be stored.

The constants in (11.3) are defined by the geometry and dielectric material of the capacitor, and can be combined to define the **capacitance** C of the particular capacitor:

$$C = \frac{A\varepsilon}{d}. \tag{11.4}$$

Then (11.3) can be rearranged to give

$$\boxed{q = Cv,} \tag{11.5}$$

which is the basic relationship between the charge on and the voltage across a capacitor. The units of C are farads (in honor of scientist Michael Faraday) with the unit symbol F. From (11.5) the unit of F can be seen to be equivalent to C/V = $C^2/(N \cdot m)$. Also, from (11.4) it can be seen that permittivity ε can be expressed in units of F/m.

In practice, the unit of F is very large (a 1.0 F capacitor is huge!), so most electronic capacitors are measured in subdivisions of F, such as nF or μF. Also, although all capacitors have two or more conducting plates facing each other as shown in the concept drawing of Fig. 11.3, to save space they are usually rolled up into cylinders or compressed by connecting many interleaved plates in parallel.

11.3   FLOW INTO AND OUT OF CAPACITORS

The symbol for a capacitor is simple: two lines close to each other representing the opposing plates of the capacitor. This is shown in Fig. 11.4. Also shown is a current i into one side (an identical current i flows out the other side) and a voltage v across the capacitor related to the stored charge by (11.5).

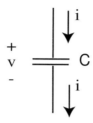

Figure 11.4: The symbol for a capacitor.

When the charge changes by an amount Δq, the voltage will also change by an amount Δv. The capacitance C is constant, so (11.5) gives

$$\Delta q = C \Delta v. \tag{11.6}$$

Now, from the definition of current in the Ohm's Law chapter [see (8.1)],

$$i = \Delta q / \Delta t. \tag{11.7}$$

Putting (11.6) into (11.7) yields

$$\boxed{\frac{\Delta v}{\Delta t} = \frac{i}{C},} \tag{11.8}$$

showing that the current into (or out of) a capacitor causes a time rate-of-change of the voltage across the capacitor, inversely proportional to the capacitance. In the limit as $\Delta t \to 0$, (11.8) becomes the differential form

$$\frac{dv}{dt} = \frac{i}{C}. \tag{11.9}$$

11.4 ANALOGY BETWEEN FLUID AND ELECTRICAL CIRCUITS

Equations (11.5) and (11.8) look familiar. We derived equations of similar form (but with obviously different variables) when we were dealing with fluids. Let's exploit this similarity between fluids and electricity. Table 11.1 is a summary of some important fluid equations in the left column, and analogous electrical equations in the right column.

With a little study of Table 11.1, we can start to pick out *pairs* of quantities (a fluid quantity and its electrical counterpart) that appear in the same place in the analogous equations. The quantities

Table 11.1: Fluid and Electrical Analogous Equations

Fluid		Electrical	
$\Delta P = QR$	Poiseuille's Law	$v = iR$	Ohm's Law
$Q = \dfrac{\Delta V}{\Delta t}$	Definition	$i = \dfrac{\Delta q}{\Delta t}$	Definition
$\dfrac{\Delta P}{\Delta t} = \dfrac{Q}{C}$	From Hooke's Law	$\dfrac{\Delta v}{\Delta t} = \dfrac{i}{C}$	From Coulomb's Law
$\tau_f = RC$	Time constant	$\tau_e = RC$	Time constant

Table 11.2: Analog Pairs

Fluid		Electrical	
resistance	R	R	resistance
compliance	C	C	capacitance
volume flow rate	Q	i	current
pressure	P	v	voltage
volume	V	q	charge
time constant	τ_f	τ_e	time constant

that pair up are called **analogs** of each other. We can then make a table of fluid/electrical analogs, Table 11.2.

Caution: Be careful not to confuse symbols. Note that Q and q are used in two different places in Table 11.2. They are used for completely *different* quantities, not even analogs of each other. The same is true of V and v. Don't get them confused. To help, lower-case symbols are used for electrical variables in this book, and upper-case symbols for fluid variables. Luckily the circuit parameters R and C have the same symbols for both fluid and electrical circuits (and resistance even has the same name).

What does the term "analog" mean in Table 11.2? Certainly the quantity in the left (fluid) column is a different physical substance from its analog in the right (electrical) column. So they are not identical quantities. But they *behave* the same, since they follow the same equations, including the difference equations. So analog pairs behave the same when their circuits are configured the same.

Therefore, it is possible—and indeed is an important engineering tool—to make an electrical circuit whose variables (various v, i, and q around the circuit) would mimic the behavior of the fluid variables (P, Q, and V, respectively) in any fluid circuit being modeled. Thus, a compliant fluid vessel is modeled by a capacitor and a resistive fluid element is modeled by a resistor. The voltage v at a given node in the electrical circuit model represents the pressure P at the corresponding node in the fluid circuit. The cardiovascular Major Project exploits this analogy to model and analyze the human CV system.

11.4.1 SCALING THE ANALOG PAIRS

But we still must deal with the translation of the resistance and compliance *values* from the fluid domain into resistance and capacitance values in the electrical domain. Is a 100 mmHg·s/L resistance in a fluid tube represented by a 100 Ω resistor? Probably not. In fact, the units are different. Remember, the analog pairs are *not* identical quantities; they only behave the same. So in general, when translating from one domain to the other, we must use a scaling factor S. To translate a fluid resistance R_f to an electrical resistance R_e, we'll use

$$R_e = S_R R_f, \tag{11.10}$$

where the factor S_R must have units such that the right-hand side of (11.10) has the units of Ω. Similarly, for capacitance:

$$C_e = S_C C_f. \tag{11.11}$$

The factor S_C must have units such that the right-hand side of (11.11) has the units of F.

Now, are we free to choose any values for the translating scale factors S_R and S_C? Not without impacting other scale factors. Note from Table 11.1 that

$$\tau_f = R_f C_f \tag{11.12}$$

and

$$\tau_e = R_e C_e. \tag{11.13}$$

If we now make the reasonable decision that we want the time scale in the electrical domain to be the same as in the fluid domain[2] (that is, that *one second* in the fluid circuit corresponds to *one second* in the electrical model), then from (11.10)–(11.13):

$$\tau_f = R_f C_f = \tau_e = R_e C_e = S_R S_C R_f C_f. \tag{11.14}$$

Canceling $R_f C_f$ terms in (11.14) gives

[2]There are some instances where it is desirable to slow down or speed up time in the analogous model compared to the system being analyzed, but we won't do that for the cardiovascular Major Project since we want the electrical model to display the same heart rate as in real life.

$$S_R S_C = 1, \tag{11.15}$$

or

$$S_R = 1/S_C. \tag{11.16}$$

Once S_C is chosen, S_R is set by (11.16). It is often the case that the electrical capacitance size is limited by the largest value available conveniently, so that will determine the scale factor S_C. Then S_R is found using (11.16).

11.5   PROBLEMS

11.1. In the circuit shown in Fig. 11.5, at time $t = 0$ the voltage $v_3 = +5.0$ V.

   a. What is the charge q on the capacitor at $t = 0$?

$$[\text{ans: } q = 5.0 \times 10^{-4} \text{ C}]$$

   b. Assuming the currents $i_1 = 2.00$ mA and $i_2 = 2.50$ mA shown in Fig. 11.5 are constant, what is the voltage v_3 on the capacitor at time $t = 400$ ms?

$$[\text{ans: } v_3 = +3.0 \text{ V}]$$

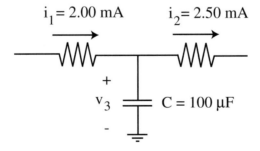

Figure 11.5: RC circuit analyzed in Problem 11.1.

11.2. The windkessel fluid circuit shown in Fig. 11.6 has the fluid element values: resistance $R_1 = 145$ mmHg·s/L, $R_2 = 505$ mmHg·s/L, and compliance $C = 0.0700$ L/mmHg.

You wish to model this circuit with an equivalent electrical circuit. To keep the resistor values from being too large, you want to use the largest capacitor value possible. The maximum capacitor value you have available is 100 μF. Determine the two scaling factors that will find the electrical values corresponding to the fluid values, one for resistors and one for capacitors,

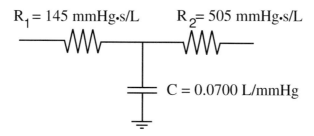

R₁ = 145 mmHg·s/L ... (values shown in figure)

Figure 11.6: Windkessel circuit to be converted in Problem 11.2.

including the correct units for the conversions. Then calculate the values for the two equivalent resistors and the capacitor.

[ans: S_R = 700 (L·Ω)/(mmHg·s), S_C = 0.00143 (F·mmHg)/L,
R_1 = 102 kΩ, R_2 = 354 kΩ, and C = 100 μF]

CHAPTER 12

Series and Parallel Combinations of Resistors and Capacitors

12.1 INTRODUCTION

We have encountered series and parallel combinations of resistive fluid elements in earlier units. We now consider series and parallel combinations of electrical resistors and of capacitors. We will initially use Kirchhoff's Laws and Ohm's Law to analyze the combinations and to reduce them to single equivalent elements, but once formulas are derived for the value of the equivalent elements, we can use these formulas directly to simplify circuits. We will see that the equations for combining electrical components are identical to those for combining analogous fluid components.

12.2 RESISTORS IN SERIES

When two or more resistors are connected in a line, as shown in Fig. 12.1 for three resistors of different values, the same current i must flow through each resistor. This is a consequence of applying Kirchhoff's Current Law (KCL) at each (nonessential) node between the individual resistors.

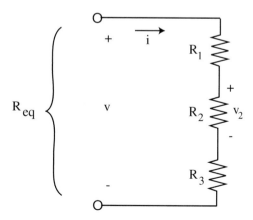

Figure 12.1: A series combination of three resistors. The equivalent resistance is R_{eq}.

Kirchhoff's Voltage Law (KVL) can be applied by taking a Kirchhoff's Tour clockwise around the loop in Fig. 12.1, using Ohm's Law directly to express the voltage drop across each resistor. This gives

$$-v + iR_1 + iR_2 + iR_3 = 0, \tag{12.1}$$

or

$$i = \frac{v}{R_1 + R_2 + R_3}. \tag{12.2}$$

Now the three resistors can be replaced by a single equivalent resistor R_{eq} (as far as the effect at the outer terminals is concerned) if the current i is the same for the equivalent resistor as for the original series combination. Thus, we require

$$i = \frac{v}{R_{eq}} \tag{12.3}$$

Comparing (12.2) and (12.3) shows

$$\boxed{R_{eq} = R_1 + R_2 + R_3} \quad \text{Series } R \text{ (for } N = 3) \tag{12.4}$$

for three resistors. In general, for N resistors,

$$\boxed{R_{eq} = \sum_{i=1}^{N} R_i} \quad \text{Series } R \text{ (any } N) \tag{12.5}$$

Note from (12.4) or (12.5) that the equivalent resistance is always *larger* than the *largest* of the individual resistors.

12.3 RESISTORS IN PARALLEL

In a parallel combination, the resistors are connected side-by-side, as shown in Fig. 12.2 for two resistors of different values. The currents i_1 and i_2 can be found by using Ohm's Law for each resistor:

$$i_1 = \frac{v}{R_1} \quad \text{and} \quad i_2 = \frac{v}{R_2}. \tag{12.6}$$

Applying KCL at node a and using (12.6),

$$i = i_1 + i_2 = \frac{v}{R_1} + \frac{v}{R_2} = v\left(\frac{1}{R_1} + \frac{1}{R_2}\right). \tag{12.7}$$

Again we require that the current i is the same when the parallel combination is replaced (across the outer terminals) by an equivalent resistance R_{eq}. Comparing (12.3) with (12.7) gives

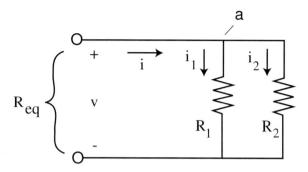

Figure 12.2: A parallel combination of two resistors having an equivalent resistance R_{eq}.

$$\frac{1}{R_{eq}} = \frac{1}{R_1} + \frac{1}{R_2} = \frac{R_1 + R_2}{R_1 R_2}. \tag{12.8}$$

Inverting (12.8),

$$\boxed{R_{eq} = \frac{R_1 R_2}{R_1 + R_2}} \quad \text{Parallel } R \text{ (for } N = 2 \text{ only)} \tag{12.9}$$

for two resistors. In general, for N resistors in parallel,

$$\boxed{\frac{1}{R_{eq}} = \sum_{i=1}^{N} \frac{1}{R_i}} \quad \text{Parallel } R \text{ (any } N) \tag{12.10}$$

Note from (12.9) or (12.10) that R_{eq} is always *smaller* than the *smallest* of the individual resistors. For example, for the simple case of two resistors of equal resistance R, (12.9) gives $R_{eq} = (1/2)R$.

Example 12.1.   Resistor Network
Find the equivalent resistance of the resistor network between the terminals marked X-X in Fig. 12.3.

Solution
First combine the three resistors that are in parallel (the lower three) using (12.10):

$$\frac{1}{R_p} = \frac{1}{50.0\,\Omega} + \frac{1}{100.0\,\Omega} + \frac{1}{70.0\,\Omega} = 0.0443/\Omega,$$

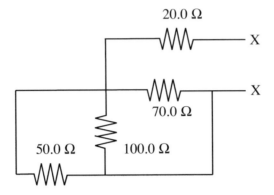

Figure 12.3: Parallel and series network of resistors analyzed in Example 12.1.

to obtain

$$R_p = 22.6 \ \Omega.$$

The network then reduces to two resistors in series (see Fig. 12.4), which can be replaced by a single R_{eq}.

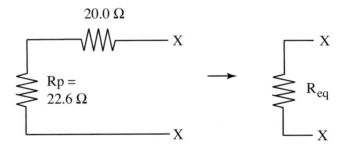

Figure 12.4: The circuit of Fig. 12.3 simplified in steps.

Using (12.5), the overall equivalent resistance is

$$R_{eq} = 20.0 \ \Omega + 22.6 \ \Omega = \mathbf{42.6 \ \Omega}.$$

12.4   CAPACITORS IN SERIES

Figure 12.5 shows two capacitors in series. Using KVL around the loop gives

$$v = v_1 + v_2. \tag{12.11}$$

Therefore, any change in the voltage across the capacitors Δv divided by the increment in time Δt is

$$\frac{\Delta v}{\Delta t} = \frac{\Delta v_1}{\Delta t} + \frac{\Delta v_2}{\Delta t}. \tag{12.12}$$

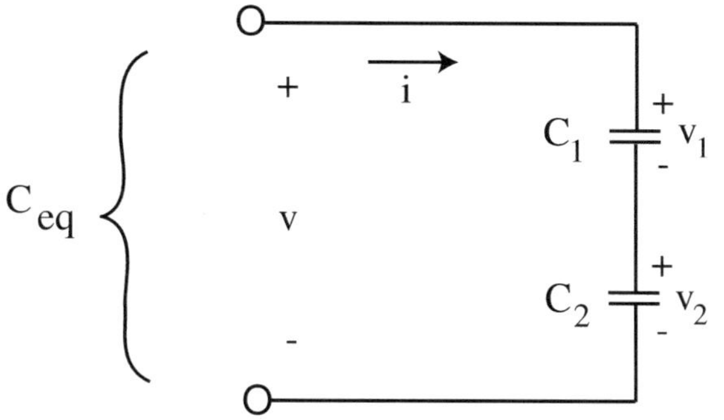

Figure 12.5: Two capacitors in series. The effect is the same as a single equivalent capacitor C_{eq}.

The relationship between the change in voltage across each capacitor and the current into the capacitor is given by (11.8) from the Coulomb's Law chapter:

$$\frac{\Delta v}{\Delta t} = \frac{i}{C}. \tag{12.13}$$

Since the current i is the same into each capacitor, using (12.13) in the right-hand side of (12.12) gives

$$\frac{\Delta v}{\Delta t} = \frac{i}{C_1} + \frac{i}{C_2} = i\left(\frac{1}{C_1} + \frac{1}{C_2}\right). \tag{12.14}$$

We can define an equivalent capacitance relating the current i to the change Δv in the overall voltage as

$$\frac{\Delta v}{\Delta t} = \frac{i}{C_{eq}}. \tag{12.15}$$

Comparing (12.15) with (12.14) shows that

$$\frac{1}{C_{eq}} = \frac{1}{C_1} + \frac{1}{C_2}, \tag{12.16}$$

or

$$\boxed{C_{eq} = \frac{C_1 C_2}{C_1 + C_2}} \quad \text{Series } C \text{ (for } N = 2 \text{ only)} \tag{12.17}$$

for the case of two capacitors in series. In general, for N capacitors in series,

$$\frac{1}{C_{eq}} = \sum_{i=1}^{N} \frac{1}{C_i} \qquad \text{Series } C \text{ (any } N) \tag{12.18}$$

Note that the formula for combining capacitors in *series* is the same as for combining resistors in *parallel*.

12.5 CAPACITORS IN PARALLEL

The combination of three capacitors in parallel is shown in Fig. 12.6. Applying KCL to node a gives

$$i = i_1 + i_2 + i_3. \tag{12.19}$$

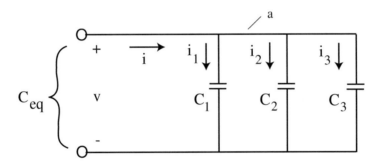

Figure 12.6: Three capacitors in parallel, giving an equivalent capacitance C_{eq}.

From (12.15), the equivalent capacitance can be found from

$$C_{eq} = \frac{i}{\left(\dfrac{\Delta v}{\Delta t}\right)}. \tag{12.20}$$

Putting (12.19) in (12.20):

$$C_{eq} = \frac{i_1}{\left(\dfrac{\Delta v}{\Delta t}\right)} + \frac{i_2}{\left(\dfrac{\Delta v}{\Delta t}\right)} + \frac{i_3}{\left(\dfrac{\Delta v}{\Delta t}\right)}. \tag{12.21}$$

Since the same voltage v appears across all capacitors, each term on the right-hand side of (12.21) is equal to [using (12.13)] the capacitance of the respective individual capacitors, giving

$$\boxed{C_{eq} = C_1 + C_2 + C_3} \qquad \text{Parallel } C \text{ (for } N = 3) \tag{12.22}$$

for three capacitors in parallel. In general, for N capacitors in parallel,

$$\boxed{C_{eq} = \sum_{i=1}^{N} C_i} \quad \text{Parallel } C \text{ (any } N) \tag{12.23}$$

Note that the formula for combining capacitors in *parallel* is the same as for combining resistors in *series*: the capacitances add. A good way to remember that capacitors in parallel *add* together, as (12.23) shows, is to recall that the equation for capacitance, (11.4), is

$$C = \frac{A\varepsilon}{d}, \tag{12.24}$$

where A is the area of the capacitor plates. Putting capacitors in parallel effectively adds their areas, so the equivalent capacitance is just the addition of the individual capacitances.

Example 12.2.   Capacitor Network
Find the equivalent capacitance of the following network at the terminals marked X-X (see Fig. 12.7).

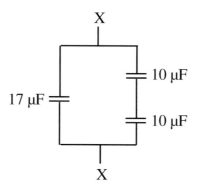

Figure 12.7: Capacitor network analyzed in Example 12.2.

Solution
First combine the two capacitors that are in series (on the right) using (12.18):

$$\frac{1}{C_s} = \frac{1}{10 \ \mu\text{F}} + \frac{1}{10 \ \mu\text{F}} = \frac{2}{10 \ \mu\text{F}} = \frac{1}{5 \ \mu\text{F}},$$

to obtain
$$C_s = 5 \ \mu\text{F}.$$

The network then reduces to two capacitors in parallel, which can be replaced by a single C_{eq} (see Fig. 12.8), and, using (12.23), the overall equivalent capacitance is

$$C_{eq} = 17 + 5 = \mathbf{22 \ \mu F}.$$

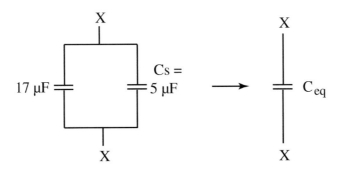

Figure 12.8: The circuit of Fig. 12.7 simplified in steps.

12.6   VOLTAGE DIVIDER

It is often desirable in electrical circuits to reduce the voltage from a source to a smaller value than the original source voltage. A circuit that will accomplish this is the **voltage divider**, shown in Fig. 12.9. Here a chain of resistors (two or more) in series is attached to the source, and the voltage across one of the resistors, say v_2, is measured.

The equivalent resistance of the series network is found from (12.4), $R_{eq} = R_1 + R_2 + R_3$, so the current i from Ohm's Law is

$$i = \frac{V_s}{R_{eq}} = \frac{V_s}{R_1 + R_2 + R_3}. \tag{12.25}$$

Using Ohm's Law again, the voltage v_2 equals $i R_2$, or

$$\boxed{v_2 = V_s \left(\frac{R_2}{R_1 + R_2 + R_3} \right)} \quad \text{Voltage divider (for } N = 3) \tag{12.26}$$

Notice that the voltage v_2 has been reduced by the fraction in the parenthesis of (12.26). This fraction is the ratio of the resistance across which the voltage of interest is measured (R_2) to the

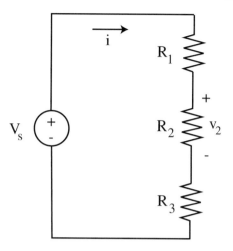

Figure 12.9: A voltage-divider network.

sum of all the resistances in series $(R_1 + R_2 + R_3)$[1]. Thus, the larger the resistor R_2, the greater the percentage of the source voltage that will be found across R_2.

12.7   CURRENT DIVIDER

When the current from a current source is split among two or more branches, this configuration is known as a **current divider**, shown in Fig. 12.10.

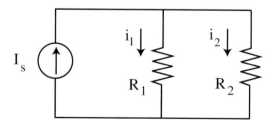

Figure 12.10: A current-divider network.

To find the common voltage v across the parallel combination of resistors, we use the equivalent resistance from (12.9) in conjunction with Ohm's Law:

---

[1]The number of resistors in the chain of the voltage divider is arbitrary. There could be two, or four, or more in series. In every case, the numerator of (12.26) remains the same, while the denominator is the sum of all the resistances in the chain.

$$v = I_s \left(\frac{R_1 R_2}{R_1 + R_2} \right).$$
(12.27)

The current through any branch, say i_2, can now be found by again using Ohm's Law:

$$\boxed{i_2 = \frac{v}{R_2} = I_s \left(\frac{R_1}{R_1 + R_2} \right)}$$ Current divider (for $N = 2$ only)   (12.28)

Note that here the fraction of I_s that passes through the branch containing R_2 is given by the ratio of the *opposite* resistor (R_1) to the sum of the resistances ($R_1 + R_2$). Thus, the larger the opposite resistor R_1, the larger the current shunted through R_2.

Example 12.3.   Voltage Divider
Using the concept of a voltage divider, find the voltage v_2 in the circuit shown in Fig. 12.11.

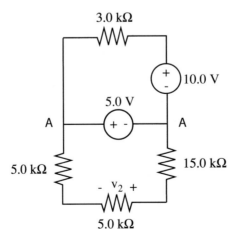

Figure 12.11: Voltage divider circuit analyzed in Example 12.3. Only the bottom portion needs to be considered in finding v_2.

Solution

The voltage source across the terminals marked A-A is an independent voltage source with a value of -5.0 V. Notice that it has a negative sign when it is referenced to the positive side of the voltage v_2. This source voltage is also across the three resistors at the bottom of the circuit. (The upper part of the circuit does not affect the voltage across A-A since the 5.0 V source is a constant voltage source.) Thus, the bottom part of the circuit forms a voltage divider. The voltage v_2 across the 5.0 kΩ resistor can be found easily using (12.26):

$$v_2 = -5.0 \text{ V} \left(\frac{5.0 \text{ k}\Omega}{5.0 \text{ k}\Omega + 5.0 \text{ k}\Omega + 15.0 \text{ k}\Omega} \right) = -5.0 \text{ V} \left(\frac{5.0}{25.0} \right) = -1.0 \text{ V}.$$

Example 12.4.   Current Divider

Using the concept of a current divider, find the current i_1 in the circuit shown in Fig. 12.12.

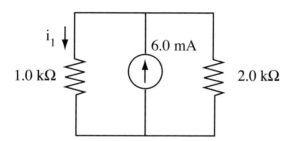

Figure 12.12: Current divider circuit analyzed in Example 12.4.

Solution

The current i_1 is flowing through the 1.0 kΩ resistor, but the *opposite* resistor has a value of 2.0 kΩ. The sum of the two resistances is 3.0 kΩ. Thus, using (12.28) and the reasoning following it,

$$i_1 = 6.0 \text{ mA} \left(\frac{2.0 \text{ k}\Omega}{1.0 \text{ k}\Omega + 2.0 \text{ k}\Omega} \right) = 6.0 \text{ mA} \left(\frac{2.0}{3.0} \right) = 4.0 \text{ mA}.$$

12.8   PROBLEMS

12.1.   a. Calculate the equivalent resistance across the terminals X-X in the circuit on the left of Fig. 12.13.

$$[\text{ans: } R_{eq} = 16.8 \text{ k}\Omega]$$

b. Calculate the equivalent capacitance across the terminals X-X in the circuit on the right of Fig. 12.13.

$$[\text{ans: } C_{eq} = 1.9 \ \mu\text{F}]$$

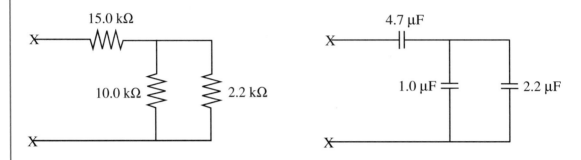

Figure 12.13: Two circuits to be reduced to their equivalents in Problem 12.1.

12.2. a. Calculate the voltage v_1 in the circuit on the left of Fig. 12.14. You don't need to do a full Kirchhoff's analysis; just use the voltage divider equation.

$$[\text{ans: } v_1 = -2.6 \ \text{V}]$$

b. Calculate the current i_1 in the circuit on the right of Fig. 12.14. You don't need to do a full Kirchhoff's analysis; just use the current divider equation.

$$[\text{ans: } i_1 = +1.2 \ \text{A}]$$

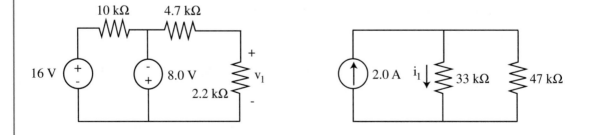

Figure 12.14: Two circuits (voltage divider on the left, current divider on the right) to be analyzed in Problem 12.2.

C H A P T E R   13

Thevenin Equivalent Circuits and First-Order (RC) Time Constants

13.1 THEVENIN EQUIVALENT CIRCUITS

It is possible, and often convenient, to transform a source from one form into another without changing its effect on the remainder of the circuit at all. One example of this is shown in Fig. 13.1. On the left is an example of a source consisting of an independent current source in parallel with a source resistance R_s (this arrangement is commonly called a Norton source) and connected to a load resistor R_L. On the right, we have transformed the source into a **Thevenin equivalent** circuit, which consists of an independent voltage source in series with the same value of source resistance R_s. The value of the Thevenin voltage source is $V_s = I_s R_s$.

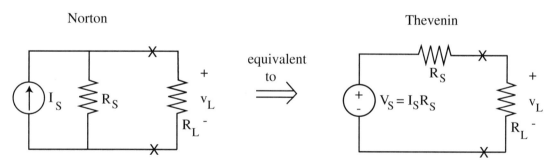

Figure 13.1: The Thevenin circuit on the right is equivalent to the Norton source on the left.

To show that this transformation has no effect on the voltage or current going to the load resistor, let's find the current coming from the Norton source (on the left in Fig. 13.1) into R_L. The load resistor and the source resistor are in parallel, so $R_{eq} = R_L R_s/(R_L + R_s)$. Since the current into the parallel combination is fixed at I_s, the voltage v_L is

$$v_L = I_s \left(R_{eq} \right) = I_s \left(\frac{R_L R_s}{R_L + R_s} \right). \tag{13.1}$$

Now consider the Thevenin equivalent circuit on the right in Fig. 13.1. The resistors are now in series, so the voltage across the load can be found from the voltage-divider formula:

$$v_L = V_s \left(\frac{R_L}{R_L + R_s} \right) = I_s \left(\frac{R_L R_s}{R_L + R_s} \right), \qquad (13.2)$$

which is the same value as found in (13.1) using the Norton source. Thus, the voltage (and current by Ohm's Law) across the load resistor is unchanged by the source transformation.

Actually, this principle is much more general. The transformation of a source into a simpler Thevenin equivalent can be done for *any* linear source, no matter how complex, and the transformation will not change the state of the remaining circuit, no matter how complex. This is shown schematically in Fig. 13.2 for any source (circuit A) and any load (circuit B).

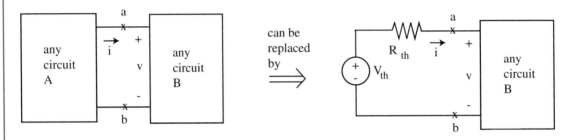

Figure 13.2: Replacing a general source by its Thevenin equivalent circuit.

Since the Thevenin equivalent circuit contains only two elements (an independent voltage source V_{th} and a series resistor R_{th}), it only takes two steps to find the Thevenin equivalent for a general circuit: first determining V_{th} and then determining R_{th}. The procedure is as follows:

Step 1. Finding V_{th} – If Thevenin's principle is true for any circuit B in Fig. 13.2, it will be true if circuit B is an open circuit, as shown in Fig. 13.3. In Fig. 13.3, no current can flow through R_{th} since it ends in an open circuit. Thus, there is no voltage drop across R_{th} for this situation, and $V_{th} = v_{oc}$. So Step 1 can be summarized:

> *Step 1 - Leave the terminals of the original circuit open and find the voltage across them. This open–circuit voltage, v_{oc} , is the Thevenin voltage; therefore $V_{th} = v_{oc}$.*

Step 2. Finding R_{th} – Here it is necessary to look back into the original circuit at its open terminals and calculate the equivalent single resistance appearing across those terminals. Before doing so, you must *deactivate* all independent sources. This means setting all independent voltage sources to a short circuit (since their voltage does not depend upon current) and setting all independent current sources to an open circuit (since their current does not depend upon voltage). Then the remaining resistances can be combined into a single equivalent resistance, which is R_{th}. So Step 2 can be summarized:

> *Step 2 - Deactivate the independent sources (see Table 13.1) and look back into the original circuit at its open terminals. Combine the resistances into one equivalent resistance R_{eq}. Then $R_{th} = R_{eq}$.*

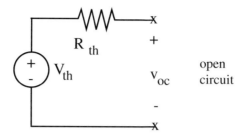

Figure 13.3: The open-circuit voltage is the same as the Thevenin voltage.

Table 13.1: Deactivating Sources
Independent voltage source → short circuit
Independent current source → open circuit

13.2   ELECTRICAL BEHAVIOR OF CAPACITORS

Resistors and capacitors behave much differently from each other with regard to their voltage and current interactions. For resistors, Ohm's Law ($v = iR$) shows a direct proportional relationship between the current through the resistor and the voltage across it. Capacitors, on the other hand, follow a more complex relationship. From (11.9) in Coulomb's Law chapter, in differential form the current-voltage relationship is

$$\frac{dv}{dt} = \frac{i}{C},$$ (13.3)

showing that the *time rate-of-change* of voltage, not the voltage directly, is proportional to the current in a capacitor. Figure 13.4 indicates the polarities of the current through a capacitor and the voltage across it.

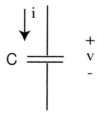

Figure 13.4: Current and voltage polarities for a capacitor.

Based upon the mathematical form of (13.3), we can investigate the validity of several possible scenarios for the waveforms of the voltage and current in a capacitor. Figure 13.5 illustrates on the upper graphs four examples of possible capacitor voltage waveforms. We ask: What must the current waveforms look like for each of these voltage waveforms in order to satisfy (13.3), and is it physically possible to have such currents?

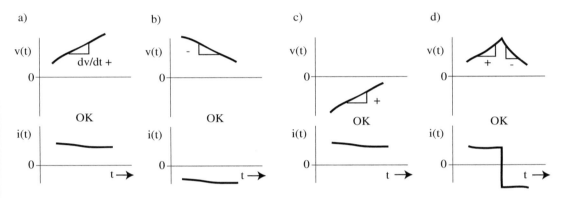

Figure 13.5: Four possible scenarios for the voltage and current waveforms across a capacitor. All of these waveforms are physically realizable.

The key for determining the current waveforms for each voltage example is to note that dv/dt is the local slope of the voltage curve. Thus, a voltage that increases with time corresponds to a positive current, while a decreasing voltage corresponds to a negative current. The currents corresponding to the various voltage waveforms are shown on the lower graphs in Fig. 13.5. All four examples have currents that are physically realizable, so all four are valid and feasible waveforms.

Figure 13.6 shows a different case. Here the voltage across the capacitor takes an instantaneous jump (a discontinuity) at one particular time. At this point, the slope of the voltage has an infinite value. Thus, the current necessary to achieve the instantaneous jump in capacitor voltage is infinite, which is course is a physical impossibility. The behavior illustrated in Fig. 13.6 is not physically realizable[1]. In other words:

Concept: The voltage across a capacitor cannot change (jump) instantaneously.

Another consequence of (13.3) is seen under conditions where a very long time has elapsed. As $t \to \infty$, the only possible value for dv/dt (the time rate-of-change of the capacitor voltage) is zero. Otherwise the voltage would keep rising to an infinite absolute value, which is not possible in the real world. Thus,

$$\text{as } t \to \infty, \ dv/dt \to 0 \tag{13.4}$$

---

[1]The only other way the voltage across a capacitor could jump instantaneously is if the capacitance value C suddenly took an instantaneous jump. But this is difficult to do physically, and is only achievable in computer simulations.

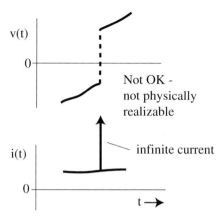

Figure 13.6: An instantaneous jump in capacitor voltage requires an infinite current, which is not physically possible.

for a capacitor. Using (13.4) in (13.3) shows that the current into the capacitor must also go to zero for long times. This means that the capacitor acts like an open circuit. In other words:

Concept: After a long time has elapsed (between changes in the source voltage), a capacitor acts like an open circuit.

The two concepts just presented will greatly facilitate the analysis of circuits containing capacitors. But note carefully: These concepts apply only to capacitors and *capacitor voltages*, not to other electrical quantities in a circuit. It is possible, for example, for the *current* into a capacitor to jump instantaneously. Also, the *voltage* across a *resistor* can change instantaneously. In fact, they often do.

13.3   RC TIME CONSTANTS

When a capacitor is connected with a resistor to a source, a simple "RC" circuit is formed, as in Fig. 13.7. If the source voltage v_1 changes abruptly (which is possible; remember that it is a *source* voltage, not the voltage across a capacitor), then the output voltage, labeled $v(t)$, changes more slowly. Being the voltage across a capacitor, $v(t)$ cannot change instantaneously. Rather it moves in an exponential manner from its initial value to some final value. Let the source voltage v_1 jump abruptly from a value of V_A to a value of V_B at some point in time, as indicated in Fig. 13.7. We will assume that the source has been at the original value V_A for a long time prior to its jump. We can also assign the time of the jump to occur at $t = 0$, since the time reference is arbitrary.

Let's analyze the circuit in Fig. 13.7 using Kirchhoff's and Ohm's Laws. Using KVL around the loop,

$$- v_1 + i R + v(t) = 0. \tag{13.5}$$

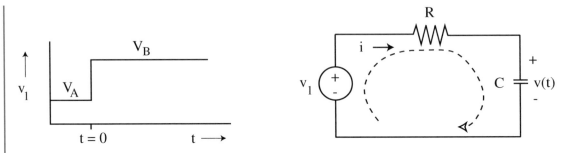

Figure 13.7: A simple RC circuit for analyzing the time behavior of the voltage across the capacitor when the source voltage v_1 changes.

(Notice that we have written the voltage across the capacitor as $v(t)$ to show explicitly that it is a function of time.) Putting (13.3) into (13.5) and rearranging results in

$$v(t) + RC\, dv/dt = v_1. \tag{13.6}$$

Equation (13.6) is a first-order differential equation describing the behavior of this circuit. You will get plenty of experience in solving equations such as this in later courses. Here we will just offer the general solution to this equation, without proof, and then use it in examples.

The general form of the solution to (13.6) is

$$\boxed{v(t) = (v_i - v_f)e^{-t/\tau} + v_f} \tag{13.7}$$

where v_i is the initial value of $v(t)$ (at $t = 0$), same as $v(0)$,

          v_f is the final value of $v(t)$ (as $t \to \infty$), same as $v(\infty)$,

and τ is the first-order time constant for voltage.

Substituting (13.7) back into (13.6) gives

$$\boxed{\tau = RC} \quad \text{Time Constant} \tag{13.8}$$

so the time constant of the circuit is given by the product of the resistance and the capacitance[2].

Let's check to make sure that the designations v_i for initial value and v_f for final value are correct in the formula of (13.7).

- Initial value: When $t = 0$, $e^{-t/\tau} = e^0 = 1$, so (13.7) becomes

$$v(0) = (v_i - v_f)1 + v_f = v_i, \tag{13.9}$$

---

[2]Note that the exponential nature of the voltage waveform is identical to that found in Chapter 6 (Part I) for the pressure in a resistive and compliant fluid system [see (6.15)]; this is not surprising since the differential equation is the same. The fluid time constant is also of identical form, being the product of the resistance and the compliance [see (6.16)].

which shows that the value of $v(t)$ at $t = 0$ is the initial value v_i, as the subscript i indicates.

• Final value: As $t \to \infty$, $e^{-t/\tau} = e^{-\infty} = 0$, so (13.7) becomes

$$v(\infty) = (v_i - v_f)0 + v_f = v_f, \qquad (13.10)$$

which shows that the value of $v(t)$ as $t \to \infty$ is the final value v_f, as the subscript f indicates.

In order to plot the waveform of (13.7) for the source voltage of Fig. 13.7, we need to find values for v_i and v_f. Here the concept of a capacitor acting like an open circuit after a long time has elapsed will help greatly.

1. First find v_i. Since the voltage v_1 has been at V_A for a long time before $t = 0$, the capacitor looks like an open circuit; thus $i = 0$ just before $t = 0$. This means that there is no voltage drop across the resistor (by Ohm's Law), which in turn means that $v(0) = V_A$. Thus, $v_i = V_A$.

2. Now find v_f. Since the final voltage is approached by waiting a very long time (actually many time constants) after the source voltage has jumped, the capacitor again looks like an open circuit. This means that $i \to 0$ as $t \to \infty$, so no voltage is dropped across the resistor and $v(\infty) = V_B$. Thus, $v_f = V_B$.

We can now plot Equation (13.7). The waveform is shown in Fig. 13.8. Note that $v(t)$ is constant at a level of V_A just before $t = 0$. Then immediately after $t = 0$, it starts rising towards its final value V_B. However, the voltage does not jump instantaneously at $t = 0$ (it can't); rather the curve forms a continuous line at $t = 0$. The rise toward V_B follows an exponential form, getting closer and closer to V_B for longer times, but the rate of rise decreases as $v(t)$ approaches V_B.

A measure of how rapidly the voltage is changing is given by the value of the RC time constant τ. At a time that is *one* time constant after the jump in the source voltage (that is, at $t = \tau$), (13.7) gives

$$\begin{aligned} v(\tau) &= (v_i - v_f)e^{-t/\tau} + v_f = (v_i - v_f)e^{-1} + v_f \\ &= v_f - v_i + v_i - v_f e^{-1} + v_i e^{-1} \\ &= v_i + (v_f - v_i)(1 - e^{-1}) \\ &= v_i + (v_f - v_i)(0.63). \end{aligned} \qquad (13.11)$$

Since v_i is the starting value of $v(t)$, and $(v_f - v_i)$ is the extent of the change (the "excursion") in $v(t)$, then (13.11) leads to the following conclusion:

At one time constant after $t = 0$, the voltage has gone 63% of its way to its final value[3].

This is shown graphically on the plot in Fig. 13.8. We will use the general formula (13.7) to solve a slightly more complex RC circuit containing two resistors in next example.

[3]After *five* time constants ($t = 5\tau$), the voltage has gone 99.3% of its way to the final value.

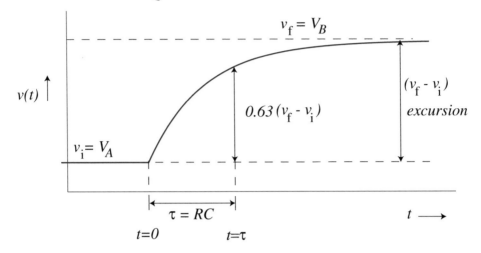

Figure 13.8: The time behavior of the capacitor voltage for an RC circuit.

Example 13.1.   Voltage across a Capacitor (and Thevenin Equivalent Resistance)
An RC circuit is shown in Fig. 13.9. The voltage v_1 jumps from $+2.00$ V to $+4.00$ V at $t = 0$. Assume it was at the level of $+2.00$ V for a long time before jumping. Find the value of $v(t)$ at $t = 200$ ms.

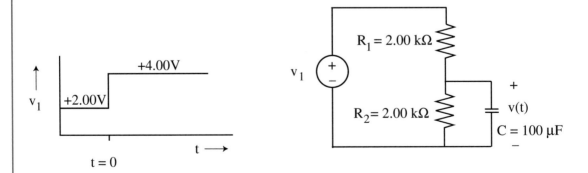

Figure 13.9: Voltage-divider circuit with a capacitor whose time behavior is analyzed in Example 13.1.

Solution

We always start by writing the general formula (13.7) for the voltage across a capacitor when the source value changes at $t = 0$:

$$v(t) = (v_i - v_f)e^{-t/\tau} + v_f. \tag{13.7}$$

The task is to find the values of v_i, v_f, and τ for the particular circuit being analyzed.

Find v_i : Since v_i has been at +2.00 V for a long time before $t = 0$, the capacitor looks like an open circuit just before $t = 0$. The circuit can be redrawn in this condition as shown in Fig. 13.10:

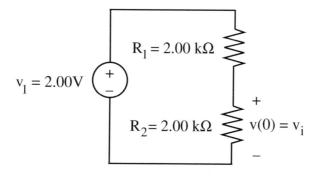

Figure 13.10: The state of the circuit just before $t = 0$.

This is a simple voltage divider. Using the voltage-divider formula,

$$v_i = v_1 \left(\frac{R_2}{R_1 + R_2} \right) = (2.00 \text{ V}) \left(\frac{2.00 \text{ k}\Omega}{4.00 \text{ k}\Omega} \right) = 1.00 \text{ V}.$$

Find v_f : A long time after $t = 0$, the capacitor will again look like an open circuit, and the overall circuit can be redrawn as shown in the figure immediately above, except that v_1 is now +4.00 V, not +2.00 V. Again using the voltage divider formula,

$$v_f = (4.00 \text{ V}) \left(\frac{2.00 \text{ k}\Omega}{4.00 \text{ k}\Omega} \right) = 2.00 \text{ V}.$$

Find τ : This is more complicated than the earlier example in Fig. 13.7 since we have *two* resistors involved for this example, R_1 and R_2. But we can use Thevenin's principle to find a single equivalent resistance at the terminals of the capacitor. We deactivate the independent voltage source (replace it with a short circuit) and look back into the circuit at the terminals of the capacitor. This situation is shown in Fig. 13.11

Note that the resistors are in parallel, so the equivalent resistance is

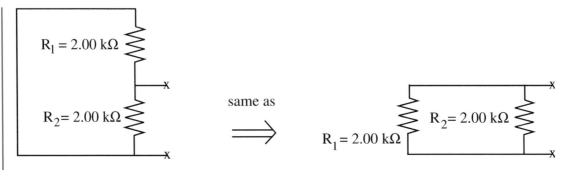

Figure 13.11: Finding the Thevenin equivalent resistance to determine τ.

$$R_{eq} = \frac{R_1 R_2}{R_1 + R_2} = \frac{(2.00 \text{ k}\Omega)(2.00 \text{ k}\Omega)}{(2.00 \text{ k}\Omega + 2.00 \text{ k}\Omega)} = 1.00 \text{ k}\Omega.$$

Then $\tau = R_{eq}C = (1.00 \text{ k}\Omega)(100 \text{ }\mu\text{F}) = 0.100 \text{ s} = 100 \text{ ms}$. We can now find v at 200 ms by substituting values into (13.7):

$$v(200 \text{ ms}) = (1.00 - 2.00) e^{-200/100} + 2.00 = 2.00 - 1.00 e^{-2.00} = 2.00 - 0.14 = \mathbf{1.86 \text{ V}}.$$

13.4 PROBLEMS

13.1. Find the Thevenin equivalent for the circuit that is to the left of the terminals X-X shown in Fig. 13.12.

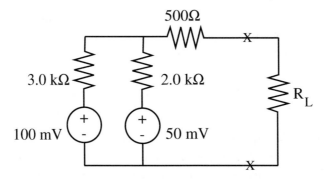

Figure 13.12: Circuit to be analyzed in Problem 13.1.

[ans: see Fig. 13.13]

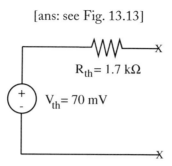

$R_{th} = 1.7\ k\Omega$

$V_{th} = 70\ mV$

Figure 13.13: Answer to Problem 13.1.

13.2. (Lab Experiment) In the lab, set up the following circuit using a prototyping board (Fig. 13.14). Apply the voltage v_1 from a function generator to match the waveform shown below. Observe the voltage v_1 using the voltage follower and one channel of an oscilloscope. Use another

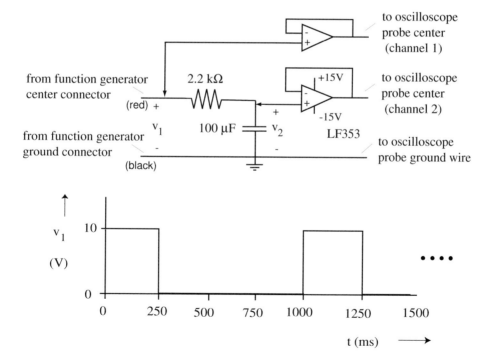

Figure 13.14: Circuit to be set up in lab for measuring voltages.

voltage follower and the second channel of the oscilloscope to simultaneously observe the voltage v_2. By **hand** and with a ruler, plot both the v_1 voltage waveform and the v_2 voltage waveform on the **same** graph axes as a function of time for a cycle or two. Label the vertical axis in units of V, but *also* label the vertical axis (that is, use a dual label) with "equivalent" blood pressure—assuming that 10 V is equivalent to 100 mmHg of pressure.

Use a sheet of green 'engineering' paper for this graph (it will be provided). Follow proper graphing techniques when doing this graph; points will be deducted from your score if proper techniques are not used. (The graph is all you need to turn in as homework for this problem, one graph per student.)

13.3. In Problem 13.2 the function generator charged a single capacitor through a resistor when its voltage v_1 was at its high value of $+10$ V, then discharged it when the voltage was 0 V. The voltage v_2 across the capacitor rose and fell cyclically.

a. Assuming the voltage v_2 at the beginning of the charging period was $+0.30$ V, calculate the voltage at the end of the charging period.

$$[\text{ans: } v_2 \text{ at } 0.250 \text{ s} = +6.9 \text{ V}]$$

b. Compare your answer in part **a** with the value you actually measured and plotted in Problem 13.2. If they are not identical, give one reason why.

13.4. The voltage v_1 in the circuit below has been $+10$ V for a long time, then drops to $+5.0$ V at $t = 0$, as shown in Fig. 13.15. Find the voltage v_2 at $t = 35$ ms.

$$[\text{ans: } v_2(35 \text{ ms}) = +2.9 \text{ V}]$$

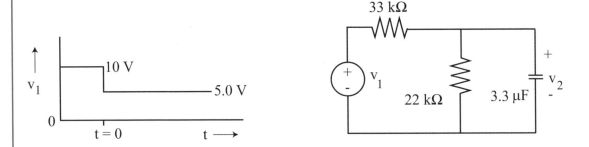

Figure 13.15: Circuit to be analyzed in Problem 13.4.

CHAPTER   14

Nernst Potential: Cell Membrane Equivalent Circuit

14.1   INTRODUCTION

We have seen in the Darcy's Law chapter how important the cell membrane is to the proper functioning of the cell in its various roles in the body. An excellent example is the neuron, the cell of the nervous system, which triggers, controls, and transmits signals to and from all parts of the body. The neuron uses action potentials—electrical impulses that travel down the length of the axon of the neuron—to transmit signals within the interconnected network of the nervous system. The ion permeability properties of the neuron produce the membrane's electrical state, and changes in that electrical state initiate and propagate the action potential. Modeling the cell membrane will employ of many of the electrical concepts covered in the previous chapters, in particular Kirchhoff's Laws (Chapter 9), the Thevenin equivalent circuit (Chapter 13), and the behavior of capacitors in circuits (Chapters 11 and 13).

14.2   CELL MEMBRANE STRUCTURE

The membrane of the neuronal cell, like that of all cells, is not an impervious barrier, as first discussed in Chapter 2 of Part I. Nutrients can enter the cell and waste products can be excreted; otherwise the cell would die. Moreover, the membrane selectively allows various ions (charged atoms or molecules) to cross its walls, setting up an electrical potential (voltage) across the membrane. This voltage, v_m, is a key element in the production of action potentials in neurons.

Figure 14.1 is a drawing of the cross-section of the cell wall of a typical neuron[1]. The microstructure of the membrane reveals two major features responsible for its selective permeability to ions and other molecules:

1. *Channels* – These are tiny water-filled pores formed by membrane-spanning proteins that extend across the thickness of the membrane, allowing movement of ions from inside to outside the cell, and vice versa. They generally are partially selective to passing a specific ion or ion type, selecting either by size or by charge. For example, there are potassium-ion (K^+) channels, sodium-ion (Na^+) channels, and chloride-ion (Cl^-) channels.
The channels, in turn, can be put into two categories:

---

[1]In drawings in this book, the membrane segment is oriented such that the region above the membrane is *inside* the cell while the region below is *outside* the cell. This is because the voltage outside the cell is usually set to be the reference (ground) voltage, and by convention, circuit diagrams usually have the ground voltage located at the bottom of the diagram (see Fig. 14.6, for example).

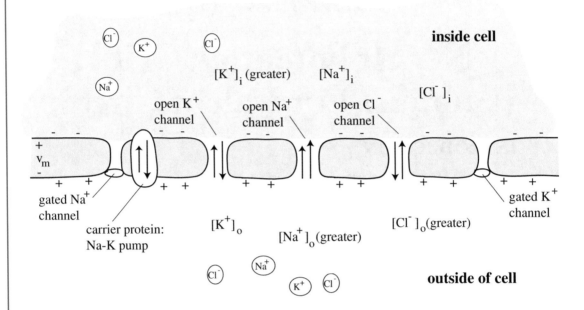

Figure 14.1: Stylized drawing of the channels (open and gated) and carrier proteins responsible for transporting ions across the cell membrane. Only a section of the neuronal cell wall is shown here.

a. Open, or passive, channels – These channels are almost always open to allow ion passage (at least for the species of ions that the channel is selective for). There is, however, some resistance that must be overcome as the ions pass through the channels.

b. Gated, or active, channels – These channels change their state (with electrical, mechanical, or chemical stimulus) from being either open or closed to the opposite, thereby modulating the flow of the specific ions. For instance, gated Na^+ channels that open when the membrane voltage reaches above a threshold value are responsible for generation of neuronal action potentials.

2. *Carrier proteins, or ion pumps* – Some of the proteins that span the membrane actively transport specific ions across the membrane barrier. These ion pumps consume metabolic energy to transport the ions against the forces that are simultaneously causing the ions to flow through open channels in the opposite direction. The carrier proteins are specific for a particular ion or pair of ions. For example, there is a Na^+-K^+ pump in the membrane of neurons that transports Na^+ ions out of the cell while at the same time transporting K^+ ions into the cell.

14.2.1 MECHANISMS OF TRANSPORT

There are three main mechanisms that can cause the transportation of ions and other molecules across membranes:

1. *Hydraulic pressure-driven transport* – We have covered this topic in the Darcy's Law chapter. Pressure-driven transport plays an important role in regulating blood volume in the capillaries and water balance in the kidneys, but it is normally not involved in the flow of ions across cell walls, and will not be considered further here.

2. *Electro-chemical diffusion-driven transport* – This is the major determinant of ion transport across cell walls. As the name implies, there are two components:

   a. *Electrically-driven flow* – Coulomb's Law describes the electrostatic force that causes like charges to repel and unlike charges to attract. If there is an excess accumulation of one species of ions on one side of the membrane, such as shown in Fig. 14.2, an electric field will be set

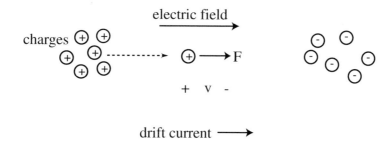

Figure 14.2: An accumulation of charge sets up an electric field, causing a drift current of ions toward the opposite charge.

   up across the membrane which exerts a force on similarly charged ions, pushing them in a direction across the membrane away from the accumulation and toward the oppositely charged ions. This flow of ions under the influence of an electric field is termed **drift current**.

   b. *Concentration-driven flow* – Due to the random motion of atoms and molecules at a finite temperature (Brownian motion), whenever there is a non-uniform concentration of any species within a volume, such as shown in Fig. 14.3 for a concentration difference across a membrane, the species tends to spread out to eventually produce a uniform concentration. Fick's Law describes the flow of atoms or molecules under the influence of a concentration gradient. It applies to charged ions as well. This flow of ions is termed **diffusion current**.

3. *Active Transport* – This mechanism describes the transport of ions or other molecules across the membrane by the carrier proteins, as discussed in the previous section. It is responsible (in combination with electro-chemical diffusion) for setting up and maintaining ionic and other concentrations

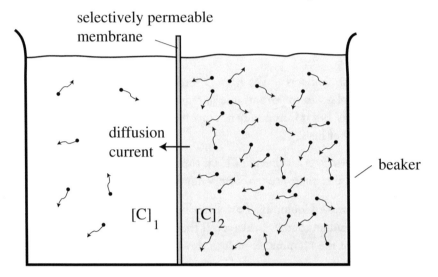

Figure 14.3: The higher concentration of C molecules on the right side of the beaker will result in a diffusion current across the selectively permeable membrane. The current is proportional to the concentration gradient (i.e., the difference between the concentration of the right side $[C]_2$ and the left side $[C]_1$).

across the membrane. Active transport requires metabolic energy, since often the movement is against voltage or concentration gradients.

14.3   NERNST POTENTIAL

Consider the situation shown in Fig. 14.4, representative of the ion distribution for a simple cell with channels selective to only one ion species. The fluid inside the cell (the intracellular fluid, or cytoplasm) is separated from the fluid outside the cell (the extracellular fluid) by the selectively permeable membrane. At least one species of ions (for example, K^+) has unequal concentrations in the two compartments—a higher concentration of K^+ inside than outside the cell, say. The open channels for K^+ allow ion flow in both directions depending on the prevailing forces. This means that there is a diffusion current flowing through the channels from the side that has the higher concentration (the inside) toward the lower concentration (the outside).

There is also a background of negatively charged ions (Cl^- if KCl was the original compound dissociated in the fluid) whose concentration almost balances the concentration of K^+ in the two regions to maintain approximate charge neutrality in each region. But there is a slight excess of Cl^- ions (not balanced by K^+ ions) in a layer immediately next to the inside of the membrane, and an equal slight excess of K^+ ions (not balanced by Cl^- ions) in a layer immediately outside

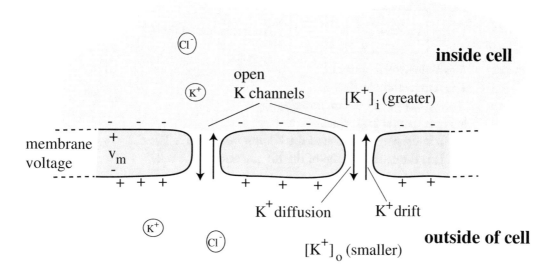

Figure 14.4: Example of a simple membrane with open K^+ channels. The concentration of K^+ inside the cell $[K^+]_i$ is larger that on the outside $[K^+]_o$ causing a diffusion current of K^+ out of the cell. However, positive charges on the outside of the membrane produce an electrical force that causes a drift current of K^+ into the cell. When the membrane voltage v_m is at the Nernst potential, these opposing currents just cancel.

the membrane. This slight imbalance is initially set up by the diffusion current described in the previous paragraph. In turn, this excess charge causes a negative voltage v_m across the membrane. The magnitude of the unbalanced charge is tiny compared to the total charge in each region (less than 1 part in 10,000), but its effect in producing v_m is major. (Note that the reference polarity of v_m is taken with respect to the outside fluid, as shown by the $+/-$ signs of v_m in Fig. 14.4. A negative v_m then means that the inside voltage is lower than the outside voltage.)

The presence of the membrane voltage means there is an electric field whose force tends to push K^+ ions from the outside to the inside of the cell. This K^+ drift current will flow through the open K^+ channels, opposing the diffusion current in the other direction. At a certain membrane voltage, called the **Nernst potential**, the two opposing currents will just cancel, so the net K^+ current is zero and the cell is in equilibrium with respect to K^+ ions. This special voltage is given (for K^+) by the formula[2]

---

[2]This formula will be derived formally later in courses in Biology, Biophysics, or Physiology. It is sometimes written in a form where the ratio in the argument of the ln function is inverted, with the outside concentration in the numerator and the inside concentration in the denominator; the negative sign then disappears.

$$\boxed{V_K = -\frac{KT}{q} \ln \frac{[\mathrm{K^+}]_i}{[\mathrm{K^+}]_o},} \qquad \text{Nernst Potential for } \mathrm{K^+} \qquad (14.1)$$

where K is Boltzmann's constant,
   T is temperature in Kelvin,
   q is the charge of the ion, *including* sign,
   ln is the natural logarithm,
   $[\mathrm{K^+}]_i$ is the concentration of the $\mathrm{K^+}$ ion inside the cell, and
   $[\mathrm{K^+}]_o$ is the concentration of the $\mathrm{K^+}$ ion outside the cell.

At room temperature, $KT/q \approx 26$ mV, so (14.1) becomes

$$V_K = -26 \ln \frac{[\mathrm{K^+}]_i}{[\mathrm{K^+}]_o} \ \mathrm{mV} \qquad (14.2)$$

at room temperature. For $\mathrm{Na^+}$ ions, the Nernst equation is

$$V_{\mathrm{Na}} = -26 \ln \frac{[\mathrm{Na^+}]_i}{[\mathrm{Na^+}]_o} \ \mathrm{mV}, \qquad (14.3)$$

and for $\mathrm{Cl^-}$ ions, the Nernst equation is

$$V_{\mathrm{Cl}} = +26 \ln \frac{[\mathrm{Cl^-}]_i}{[\mathrm{Cl^-}]_o} \ \mathrm{mV} \qquad (14.4)$$

where the positive sign is due to the negative charge of the chloride ion.

It is important to note that the Nernst potential for any ion depends on the two concentrations of the ion across the membrane, and will change if the concentration changes. Also, the Nernst potential is *not* usually the actual voltage v_m across the membrane, especially when there are two or more ion species for which the membrane is permeable, as will be seen shortly. This is because v_m is influenced by other factors in addition to the Nernst voltages, including the permeability values for each ion species and the existence of ion pumps in the membrane. Examples of finding the actual membrane voltage using the membrane equivalent circuit are given in the next section.

14.4   EQUIVALENT CIRCUIT FOR THE MEMBRANE

For a simple membrane such as shown in Fig. 14.4 that has only channels for *one* species of ions, an equivalent circuit can be constructed like Fig. 14.5. It consists of an independent voltage source of Nernst potential magnitude, V_K, in series with a resistor R_K representing the electrical resistance of the channels to $\mathrm{K^+}$-ion flow. The top terminal is at the potential of the intracellular fluid, while the bottom terminal is at the potential of the extracellular fluid, usually taken at zero voltage as the reference.

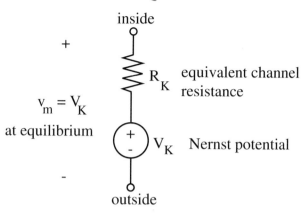

Figure 14.5: Equivalent circuit for the ion flow across a simple membrane (such as Fig. 14.4) with channels permeable only to K^+ ions.

Since K^+-ion flow provides the only current through this simple membrane, there are no other branches to its equivalent circuit. When the net current is zero (at equilibrium), there is no voltage drop across the resistor and the membrane voltage v_m is equal to the Nernst potential V_K. This single-ion equilibrium situation is one of the few cases where the membrane potential is equal to the Nernst voltage.

For a more general membrane, such as was shown earlier in Fig. 14.1 for a neuron, the equivalent circuit is more complicated. Each ion for which the membrane has channels is represented by a separate branch with its own Nernst voltage source in series with a unique resistor, as shown in Fig. 14.6. The upper wire represents the cytoplasm inside the cell (which has relatively high conductivity compared to the channels and whose resistance can therefore be ignored), and the lower wire represents the extracellular fluid (which also has high conductivity)[3].

Although the $+/-$ polarity of all the sources shown in Fig. 14.6 is oriented in a direction such that the reference voltage is at the bottom (i.e., outside the cell), individual Nernst voltages may have negative values depending upon the particular concentration differences and sign of the ion charge. Also, the values of the resistances can, and will, change as the respective gated channels open and close. In addition, to represent the capacitance of the membrane with its opposite charges separated across a barrier, a capacitor C_m has been added to the equivalent circuit in Fig. 14.6.

For the squid giant axon, a much-studied neuron, the values of the ion concentrations and equivalent resistances are listed in Table 14.1. The concentration values can be used to find the Nernst potentials for each ion using (14.2)–(14.4), and these are also listed in Table 14.1.

---

[3]The simple equivalent circuit shown in Fig. 14.6 assumes that the ion current through each branch is linearly related to the voltage across that branch, which is approximately correct for the channels in a squid giant axon membrane. Other cells are better described by a slightly more complex, and nonlinear, model defined by the Goldman-Hodgkin-Katz (or simply the Goldman) equation. However, the qualitative behavior of the ion currents and membrane voltage is similar in both models.

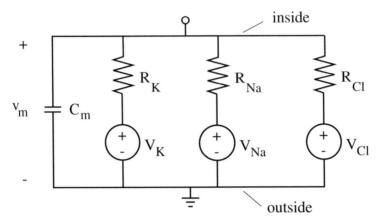

Figure 14.6: Equivalent circuit for the membrane of a typical neuronal cell, as illustrated in Fig. 14.1, which is permeable to three ions. Also included is a capacitor representing the capacitance of the membrane.

Table 14.1: Values for a Squid Giant Axon at Rest					
	$[conc]_i$	$[conc]_o$	R	V_{Nernst}	C_m
K^+	397 mM	20 mM	2.7 kΩ	−78 mV	
Na^+	50 mM	437 mM	30.0 kΩ	+56 mV	1.0 μF
Cl^-	40 mM	556 mM	3.3 kΩ	−68 mV	

The values from Table 14.1 can then be put into the equivalent circuit, as done in Fig. 14.7. A study of this figure shows that for the general case, the membrane voltage is *not* the same as any of the Nernst potentials, since in general current must be flowing through the resistors, even at equilibrium. The actual membrane voltage is found by a circuit analysis of the complete circuit; this will be done in an example next.

But first some observations can be made here: The magnitudes of the Nernst voltage sources are all different, and in fact, some have opposite signs. The overall membrane voltage will be determined, therefore, by the relative magnitudes of the series resistors. The smaller the resistance, the closer the membrane voltage will be to the Nernst voltage of that branch; the larger the resistance, the more isolated the membrane voltage is from the Nernst voltage of that branch. Since in Fig. 14.7 the K^+ branch and the Cl^- branch have much smaller resistances (corresponding to larger permeabilities) than the Na^+ branch, the membrane voltage will be negative, close to (but not quite as negative as) the Nernst voltage for K^+ and for Cl^-. In fact, as will be seen in the example next, the resting membrane voltage for the squid giant axon is about −67 mV.

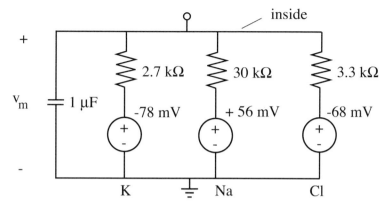

Figure 14.7: Equivalent circuit with three ion channels for the squid giant axon at rest.

Also, a careful study of the equilibrium condition shown in Fig. 14.7 reveals a seeming inconsistency. At equilibrium, no net current flows out of the top of the complete circuit. This means that the current flowing down the K^+ branch, composed of K^+ ions flowing from inside to outside the cell, must equal the current flowing up in the Na^+ branch, composed of Na^+ ions flowing from outside to inside the cell. (Note: It will be seen shortly that the Cl^- current is nearly zero.) But these two currents carry *different* ions, and after a while, if left alone, these currents would lead to changes in the respective concentrations inside and outside the cell, a violation of the equilibrium condition. The answer to this enigma is that there is an active Na-K pump across the membrane—not shown for simplicity in the equivalent circuit but illustrated in Fig. 14.1—that pumps K^+ ions into the cell and Na^+ ions out of the cell, thus maintaining the concentrations at their equilibrium values.

Finally, a note about units. In practice, the magnitude of the membrane branch resistances for the total cell are often given in terms of conductance G, which is the inverse of the resistance R; so $G = 1/R$, in units of 1/ohm=siemens (S). Even more, the membrane conductance is usually specified *per unit surface area* of the membrane, with the symbol g, in typical units of mS/cm^2. Thus, $G = g \cdot A_m$, where A_m is the total surface area of the membrane under consideration. (Similarly, the resistance R may be normalized to unit surface area using the symbol r, with typical units of k$\Omega \cdot$ cm^2. In this case, $r = 1/g$ and $R = r/A_m$.) The capacitance C_m is also often given as a capacitance per unit area c_m; most cell membranes have a capacitance of about 1 μF/cm^2.

Then if the branch conductance g (or resistance r) and capacitance c_m values are specified per unit surface area, they can be inserted into the equivalent circuit of Fig. 14.7 in one of two ways:

- Each g and c_m value can be adjusted by multiplying it by the total surface area of the membrane A_m, and the resulting conductance G inverted to give the resistance $R = 1/G$ before R and C_m are put in the circuit; or

- The $r = 1/g$ and c_m values can be inserted directly as they stand into the equivalent circuit, in which case the equivalent circuit represents a unit area of the membrane (and the branch currents are then found per unit area). This latter case is illustrated in a homework problem at the end of this chapter.

In either case the resulting membrane voltage and membrane time constant are the same.

Example 14.1.   Resting Potential Across Cell Membrane

For the membrane properties of a squid giant axon given in Table 14.1:
a. Calculate the Nernst potential for each ion.
b. Calculate the resting membrane voltage using only the K^+ and Na^+ branches.
c. Find the RC time constant of the membrane.

Solution
a. Equations (14.2)–(14.4) can be used to find the Nernst potential at room temperature. Straightforward math gives the values shown in Table 14.1. The equivalent circuit is then given by Fig. 14.7.
b. To find the resting membrane potential, we redraw the equivalent circuit of Fig. 14.7 using only the K^+ and Na^+ branches. (The reason why the Cl^- branch is not included here will be explained shortly.) The capacitor is not included since it acts like an open circuit in the steady-state or resting condition. The redrawn circuit is shown in Fig. 14.8.

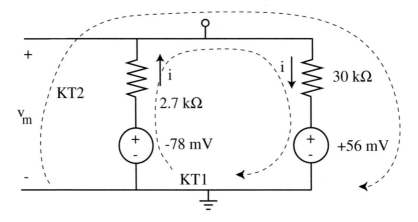

Figure 14.8: Equivalent circuit of squid giant axon membrane to find the resting membrane voltage.

The branch-current method can then be used to solve for v_m. Using KVL around the loop labeled KT1 gives

$$- (-78 \text{ mV}) + i(2.7 \text{ k}\Omega) + i(30 \text{ k}\Omega) + 56 \text{ mV} = 0, \tag{14.5}$$

so

$$i = (-134 \text{ mV})/32.7 \text{ k}\Omega = -4.1 \ \mu\text{A}. \tag{14.6}$$

Using KVL around the loop labeled KT2 gives

$$- v_m + i(30 \text{ k}\Omega) + 56 \text{ mV} = 0. \tag{14.7}$$

Putting the value for i from (14.6) into (14.7) gives the resting membrane voltage:

$$v_m = 56 \text{ mV} + (-4.1 \ \mu\text{A})(30 \text{ k}\Omega) = 56 \text{ mV} - 123 \text{ mV} = \mathbf{-67mV}. \tag{14.8}$$

Note that the value of the resting membrane voltage is very close to the Nernst potential for the Cl^- ions as given in Table 14.1. This means that there is essentially no current flowing through the Cl^- branch, and it has little effect on the remainder of the circuit. It is therefore valid to exclude it from the equivalent circuit when calculating the membrane voltage. The basic reason for this is that Cl^- is passively distributed across the membrane by open channels only with no ion pump for Cl^- involved, and the Cl^- concentrations on both sides therefore adjust themselves to match the membrane voltage set by the K^+ and Na^+ concentrations.

c. The time constant is given by the product of $R_{eq}C_m$, where C_m is the capacitance (1.0 μF) of the membrane and R_{eq} is the equivalent resistance of the membrane at rest. This resistance is found by calculating the Thevenin equivalent resistance of the circuit as seen from the terminals of the capacitor. The circuit includes all branches of the equivalent circuit. After disabling the independent voltage sources (replacing with shorts), the circuit looks like Fig. 14.9.

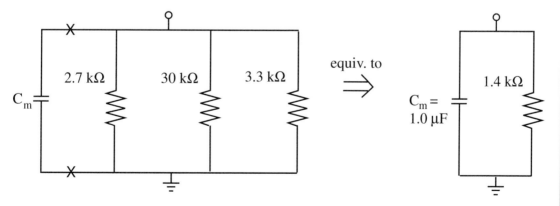

Figure 14.9: Finding the Thevenin equivalent resistance to determine the membrane time constant.

The equivalent resistance is the parallel combination of the three resistors, so

$$1/R_{eq} = 1/2.7 \text{ k}\Omega + 1/30 \text{ k}\Omega + 1/3.3 \text{ k}\Omega. \tag{14.9}$$

Thus,

$$R_{eq} = 1.4 \text{ k}\Omega, \tag{14.10}$$

and

$$\tau = R_{eq}C_m = \mathbf{1.4 \text{ ms}}. \tag{14.11}$$

Note: If it remained constant, this RC time constant of 1.4 ms would be problematic (as shown in Chapter 13 and Fig. 13.8), since the voltage across the membrane capacitance, v_m, cannot change at a rate faster than the time constant. In fact, it reaches only 63% of its excursion during one time constant and thus takes several time constants (perhaps 5 to 6 ms for a time constant of 1.4 ms) to accomplish an appreciable change. Yet nerve action potentials require a much faster rate of rise and fall than this; the complete rise and fall phases of the membrane voltage that comprise an action potential take less than about 3 ms. Luckily, as will be seen in the next example, the RC time constant of the membrane decreases significantly during the course of an action potential.

14.5   ACTION POTENTIALS

The situation shown in Table 14.1 and Fig. 14.7 holds for a resting axon only. When the cell is resting, the gated K^+ and Na^+ channels are in their resting position, having not been stimulated to change their state. But under stimulation through its synapse with neighboring neurons, the cell's membrane voltage and the state of the gated channels change dramatically. When the membrane voltage is raised above a threshold (about -40 mV for the squid giant axon) by stimulation from adjacent cells, many gated Na^+ channels open rapidly with a major effect. The resistance of the Na^+ branch declines precipitously to a very low value, much lower than the resistance of the K^+ branch. This rapidly moves the membrane voltage to a value close to the Na^+ Nernst voltage, which is positive at $+56$ mV.

The gated K^+ channels also begin to open, but at a slower rate than the Na^+ channels. This delayed action moves the membrane voltage back from its positive value again toward the K^+ Nernst voltage at -78 mV. The overall result is a voltage spike in the membrane voltage, called an **action potential**, which begins at the resting voltage, rises rapidly to a positive voltage as the Na^+ channels open, then declines back toward the resting voltage (actually overshooting in the negative direction) as the K^+ channels open and the Na^+ channels close again. The K^+ channels then close shortly after, returning the cell to its resting state ready for the next stimulation.

The situation leads to action potentials as plotted in Fig. 14.10. The upper graph shows the time behavior of the conductance G of both the K^+ and Na^+ channels just after the stimulus, when the membrane voltage rises above threshold. (It is easier to graph conductance G than resistance R; as discussed, conductance is the inverse of resistance, or $G = 1/R$.) As seen in the figure, the conductance of the Na^+ channels increases rapidly, causing the membrane voltage—shown in the lower graph—to rise abruptly toward the positive Na^+ Nernst potential. The rise in K^+ conductance is delayed slightly, so by the time it starts to increase, the Na^+ conductance is decreasing. This causes the membrane voltage to fall back toward the negative K^+ Nernst potential, producing the declining

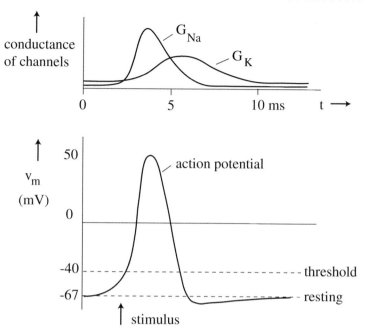

Figure 14.10: When the gated Na^+ channels rapidly open as the membrane voltage crosses the threshold value in response to a stimulus, the Na^+ conductance G_{Na} rises (upper graph) and the membrane voltage rises (lower graph) toward the positive Na^+ Nernst potential, starting the action potential. It falls when the Na^+ conductance G_{Na} falls and the K^+ conductance G_K rises, slightly later.

segment of the action potential. The increased K^+ conductance during this phase actually results in a more negative voltage than at rest, so the declining segment dips below the resting value. It slowly returns to the resting potential when the K^+ conductance declines to its resting value. This completes the cycle of the action potential, and the nerve is in a state to again be stimulated.

The generation of the action potential, with its accompanying transmission down the nerve axon, is the major electrical event responsible for the signaling and control functions of the nervous system. As we have seen, it is made possible by the unique ion permeability characteristics of the neuronal cell membrane.

Example 14.2.   Rising Phase of the Action Potential
For the squid giant axon analyzed in the previous example:
a. Assume that the Na^+ channels now open up rapidly upon stimulation such that the Na^+ conductance rises to $G_{Na} = 0.12$. Using the equivalent circuit from the previous example with a new

resistance value for R_{Na}, calculate the voltage that the membrane potential will rise toward (the peak of the action potential if all else were constant).

b. Calculate the modified time constant of the membrane at this peak of the action potential.

Solution

a. The voltage that the membrane potential will rise toward can be found by calculating the equilibrium voltage with $R_{Na} = 1/G_{Na} = 1/0.12 = 8.3\ \Omega$. The equivalent circuit becomes as shown in Fig. 14.11.

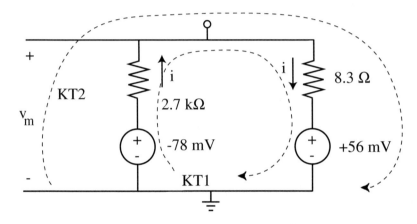

Figure 14.11: Equivalent circuit for estimating the peak voltage of an action potential.

Following steps similar to before, KVL around loop KT1 gives

$$-(-78\ \text{mV}) + i(2.7\ \text{k}\Omega) + i(8.3\ \Omega) + 56\ \text{mV} = 0, \tag{14.12}$$

so

$$i = (-134\ \text{mV})/2.7083\ \text{k}\Omega = -49.5\ \mu\text{A}. \tag{14.13}$$

Then KVL around loop KT2 gives

$$-v_m + i(8.3\ \Omega) + 56\ \text{mV} = 0, \tag{14.14}$$

or

$$v_m = 56\ \text{mV} + (-49.5\ \mu\text{A})(8.3\ \Omega) = 56\ \text{mV} - 0.4\ \text{mV} = \mathbf{+55.6}\ \textbf{mV}, \tag{14.15}$$

which is very close to the Na^+ Nernst voltage of $+56$ mV. The peak voltage of the action potential comes close to this value (rising to about $+50$ mV) before other changes (a subsequent decrease in G_{Na} and an increase in G_K) cause it to drop back again, completing the action potential spike.

b. The new time constant is much shorter than its resting value due to the reduced resistance of the membrane. The new Thevenin equivalent resistance is given by

$$1/R_{eq} = 1/2.7 \text{ k}\Omega + 1/8.3 \, \Omega + 1/3.3 \text{ k}\Omega, \tag{14.16}$$

so

$$R_{eq} = 8.25 \, \Omega, \tag{14.17}$$

and

$$\tau = R_{eq} C_m = (8.25 \, \Omega)(1.0 \times 10^{-6} F) = \mathbf{8.25 \, \mu s}. \tag{14.18}$$

This time constant (reached when G_{Na} is at its maximum value) is now short enough not to restrict the rapid rise and fall of the action potential voltage, whose timing is really determined by the changes in ion conductance.

   In obtaining the peak voltage during the action potential in the previous example, it was assumed that the *Nernst* voltages for Na^+ and K^+ did not change during the course of the action potential. But when the gated Na^+ channels open up (and the delayed K^+ channels also), ions of both species will cross the membrane at rates greater than at rest, potentially changing their concentrations inside and outside the cell, thereby upsetting the Nernst potentials, which are based upon these concentration values. It is shown next, however, that the relative concentration changes that occur during an action potential are very small compared to the resting concentrations.

   We first find the number of ions that cross the membrane during a typical action potential rise where the initial membrane voltage is -67 mV (resting value) and the peak voltage is $+50$ mV. The voltage change is $\Delta v = 50$ mV $- (-67$ mV$) = 117$ mV. According to Eq. (11.6) and using $C = 1.0 \, \mu$F for the squid giant axon, the change in the charge stored in the capacitor of the membrane is

$$\Delta q = C \Delta v = (1.0 \times 10^{-6} \text{ F})(117 \times 10^{-3} \text{ V}) = 1.2 \times 10^{-7} \text{ C}. \tag{14.19}$$

Since the charge of each ion is $q = 1.6 \times 10^{-19}$ C, this means that $(1.2 \times 10^{-7})/(1.6 \times 10^{-19}) = 7.5 \times 10^{11}$ ions cross the membrane. This seems like a large number, except when compared to the number of ions already in solution. Assuming an equilibrium concentration of Na^+ ions, for example, of 50 mM, a squid giant axon cell volume of about 1×10^{-5} L, and using the Avogadro constant, the number of Na^+ ions inside the cell at equilibrium is

$$\begin{aligned} N = & \text{ (molarity) (Avogadro constant) (volume in L)} \\ = & (50 \times 10^{-3})(6.02 \times 10^{23})(1 \times 10^{-5}) = 3.0 \times 10^{17}. \end{aligned} \tag{14.20}$$

   Thus, the relative change is concentration is only $(7.5 \times 10^{11})/(3.0 \times 10^{17}) = 2.5 \times 10^{-6}$, or about 3 parts per million, or 0.0003%, which is insignificant.

14.6   PROBLEMS

14.1. The table below gives the characteristics of a typical frog muscle cell. Note that the ion conductance and membrane capacitance values are given *per unit area*:

<table>
<tr><td colspan="6" align="center">Values for Frog Skeletal Muscle at Rest</td></tr>
<tr><td></td><td>$[conc]_i$</td><td>$[conc]_o$</td><td>g</td><td>V_{Nernst}</td><td>c_m</td></tr>
<tr><td>K^+</td><td>140 mM</td><td>2.5 mM</td><td>0.21 mS/cm^2</td><td></td><td></td></tr>
<tr><td>Na^+</td><td>13 mM</td><td>110 mM</td><td>0.023 mS/cm^2</td><td></td><td>2.1 μF/cm^2</td></tr>
<tr><td>Cl^-</td><td>3.0 mM</td><td>90 mM</td><td>0.11 mS/cm^2</td><td></td><td></td></tr>
</table>

a. Calculate the Nernst potentials at room temperature for each of the ions, and *draw the equivalent circuit* for the cell membrane at rest. Let this equivalent circuit represent a unit area of the membrane surface (i.e., 1 cm^2). Remember that conductance g is the inverse of resistance r, so $r = 1/g$.

$$[\text{ ans: } V_K = -105 \text{ mV}, \quad V_{Na} = +56 \text{ mV}, \quad V_{Cl} = -88 \text{ mV}]$$

b. Using the equivalent circuit with only the K^+ and Na^+ branches, calculate the membrane voltage at rest.

$$[\text{ans: } v_m = -89 \text{ mV}]$$

c. Calculate the RC time constant of the membrane at rest.

$$[\text{ans: } \tau = 6.1 \text{ ms}]$$

14.2. *Muscle Action Potential (MAP)*

a. Now assume that the gated Na^+ channels in the muscle cell of Problem 14.1 quickly open up such that the overall conductance g_{Na} of the Na^+ channels increases rapidly to 20 mS/cm^2. Calculate the resulting membrane voltage. (This represents the voltage toward which the action potential is rising. Use the equivalent circuit from Problem 14.1 again with only the Na^+ and K^+ branches but with the new r for the Na^+ branch.)

$$[\text{ans: } r_{Na} = 50 \ \Omega \cdot cm^2, \quad v_m = +54 \text{ mV}]$$

b. Calculate the new RC time constant of the membrane under the condition of the new r for the opened Na^+ channels.

$$[\text{ans: } \tau = 0.10 \text{ ms}]$$

c. In 1-2 sentences, explain why is it essential for the membrane time constant found in this problem to be so much lower than that of Problem 14.1 (at rest) in order to allow a muscle action potential spike.

CHAPTER 15

Fourier Transforms: Alternating Currents (AC) and the Frequency Domain

15.1 INTRODUCTION - SINUSOIDS

Many of the voltages we have encountered so far have been constant, or nearly constant. An example is a battery modeled as an independent voltage source. Such steady sources are called **dc** signals, for "direct current" (even though they may be voltages or pressures, not strictly currents). But some important signals are time-varying, such as the aortic pressure pulse and the action potential in neurons. It is useful to describe these time-varying signals not only by their time waveforms but by a complementary viewpoint: their frequency content. Signals that vary in time are called **ac** signals, for "alternating current."

A very important and common time-varying waveform is the **sinusoid**. A voltage that has a sinusoidal time behavior can be written in general as[1]

$$v = V_0 \cos(2\pi f t - \Phi), \qquad (15.1)$$

where cos is the cosine function,

V_0 is the peak voltage (V),

f is the frequency (hertz, or Hz, equivalent to 1/s),

t is time (s), and

Φ is a phase angle (radians or degrees, dimensionless).

Figure 15.1 plots the sinusoidal waveform of (15.1) in two ways: (a) as a function of the argument of the cosine term, and (b) as a function of time. The phase angle Φ represents the offset between the peak of the cosine wave and the $t = 0$ origin, as shown in Fig. 15.1(a). For the special case when $\Phi = \pi/2$ (or 90°), the cosine wave can be written as

$$v = V_0 \cos(2\pi f t - \pi/2) = V_0 \sin(2\pi f t). \qquad (15.2)$$

In the time plot, the sinusoid repeats with period T. This occurs whenever its argument increases by 2π. So from (15.1), the relationship between period T and frequency f is $2\pi f T = 2\pi$, or

$$\boxed{f = 1/T.} \qquad (15.3)$$

---

[1] General sinusoidal waves, like (15.1), are usually called "sine" waves for short, even though they may be written in terms of cosines.

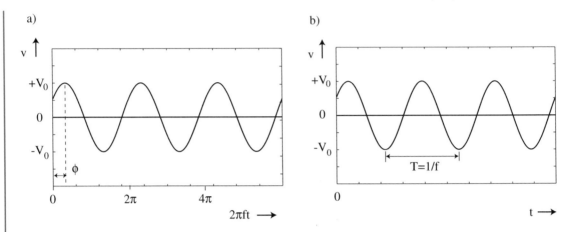

Figure 15.1: Plots of the sinusoidal wave given by (15.1): (a) as a function of the argument of the cosine term, and (b) as a function of time t.

The reason sinusoidal waves are so important in engineering analysis is that the natural response of second-order systems (a mass on a spring, or a combined capacitor-inductor electrical circuit, for example) follows a sinusoidal behavior in time. Thus, many physical phenomena, including the displacement of a vibrating string and the pressure of ultrasound waves in tissue, can be described by sine waves at discrete frequencies.

15.2   FOURIER SERIES FOR ARBITRARY REPEATING WAVEFORMS

In fact, sine waves at all possible frequencies form a "complete set." This means that *any*[2] arbitrary waveform (a pulse, a square wave, a triangular wave) can be written as a combination of sine waves of various amplitudes and phases. If the arbitrary waveform is one-time only, a continuous spectrum of frequencies is needed to represent the signal. However, if the waveform is repetitive, sine waves of only certain discrete frequencies are needed. The lowest of these components has a frequency of zero, or $f_0 = 0$, which is the **dc** or steady component. The next lowest frequency is the **fundamental** frequency, with a value of

$$f_1 = 1/T, \tag{15.4}$$

where T is the repeat period of the waveform. In addition, a series of higher frequencies, called **harmonics**, which are integral multiples of the fundamental frequency, are required to completely describe the waveform. The harmonic frequencies are given by

[2]There are some very unusual waveforms—those with an infinite number of infinite discontinuities—that cannot be represented by a Fourier series, but these are not of interest to us.

$$\boxed{f_n = nf_1 = n/T} \qquad \text{for } n = 0,1,2.... \qquad (15.5)$$

where n is the harmonic number. For example, f_2 is the second harmonic, f_5 is the fifth harmonic, etc. Note that (15.5) also gives the correct frequency for the dc component ($n = 0$) and for the fundamental ($n = 1$).

Mathematically, any arbitrary repetitive signal $g(t)$ with fundamental period T can be written as a **Fourier series**:

$$g(t) = \sum_{n=0}^{\infty} A_n \cos\left(2\pi f_n t - \phi_n\right), \qquad (15.6)$$

where A_n is the amplitude of the nth harmonic component,
  Φ_n is the phase of the nth harmonic component, and
  f_n is the harmonic frequency given by (15.5).
The individual terms in the series (15.6) are called the Fourier components. The "sharper" the features of the signal, the larger the amplitudes A_n of the higher harmonics making up the waveform. For example, a sharp spike contains a large number of significant high-frequency Fourier components. On the other hand, a smooth single sinusoidal wave has only one component: the fundamental.

As a simple example of a repetitive waveform, consider the rectangular waveform shown in Fig. 15.2 as a function of time (that is, in the "time domain"). Its "on" time is 250 ms, its "off" time

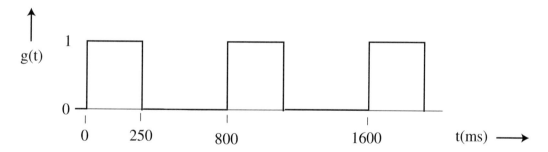

Figure 15.2: A repetitive rectangular waveform.

is 550 ms, and it repeats with a period of 800 ms. Its Fourier frequency components are shown in Fig. 15.3 (in the "frequency domain"). Both the amplitude A_n and phase Φ_n are plotted for each frequency component. Note that the fundamental frequency is $1/T = 1/(800 \text{ ms}) = 1.25$ Hz, and that the higher harmonics are integral multiples of 1.25 Hz. The amplitude of the dc term is just the *average* value of the signal, which can be calculated in another way for a rectangular waveform by finding the peak value multiplied by the ratio of the "on" time to the period, or $1 \times (250 \text{ ms}/800 \text{ ms}) = 0.3125$.

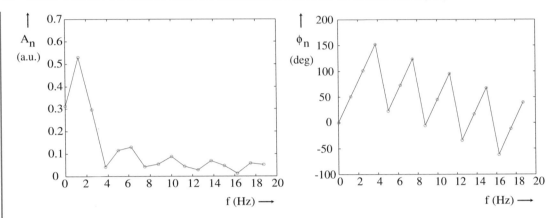

Figure 15.3: The amplitude A_n and phase ϕ_n of the Fourier components of the rectangular wave shown in Fig. 15.2.

It is important to realize that Figs. 15.2 and 15.3 are just two different ways of representing the *same* signal. Both contain all the information inherent in the signal. In fact, the time waveform—Fig. 15.2—can be reconstructed by adding together all its Fourier terms using the amplitudes and phases from Fig. 15.3 for all frequency components. The result of this progressive summation at different stages (for increasing numbers S of terms added in) is shown in Fig. 15.4, starting with the first two terms ($S = 2$, for $n = 0$ and $n = 1$), and ending with all the terms ($S = 16$) added together.

It's evident from the $S = 2$ curve in Fig. 15.4 that the fundamental term sets the overall period of the wave, and the higher harmonics represent the sharper details of the wave. As the higher harmonics are added (see $S = 16$), the summation approaches the original waveform.

15.3   FFT: CALCULATING THE DISCRETE FOURIER TRANSFORM

There are several formulas available for finding the frequency components of a general time waveform. Mathematically they involve integrals over the time variable of the product of the waveform and sine waves. You will use one or more of these mathematical formulas in later courses (math and engineering classes), but for this book we will use the computer to find the frequency components, in particular by using Matlab as our programming and analysis tool.

The algorithm that most software programs employ to find frequency components is the **FFT** algorithm, which stands for "Fast Fourier Transform." To use this method, the time waveform first needs to be sampled at discrete, uniformly spaced time intervals. According to the **Nyquist criterion**, the sampling interval Δt should be small enough that there are at least *two* samples per period T_h of the *highest* frequency component contained in the original waveform (we'll see why shortly), or $\Delta t \leq T_h/2$. Since $T = 1/f$, the Nyquist criterion requires that

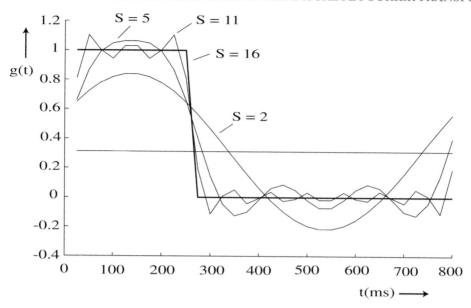

Figure 15.4: The recombination of the Fourier series for the rectangular wave at different stages. Shown are the results for adding 2 terms, 5 terms, 11 terms, and 16 terms (the maximum), respectively.

$$\boxed{\Delta t \leq 1/(2 f_h)} \qquad \text{Nyquist Criterion} \qquad (15.7)$$

where f_h is the highest frequency contained in the waveform.

As an example of the use of the FFT algorithm, Fig. 15.5 shows the waveform of a typical repeating action potential (discussed earlier in the Nernst Potential chapter) for one cycle. Plotted on the continuous voltage curve are the points where the voltage value has been *sampled*, at 16 points in this example. The array of these values (a row vector in Matlab language) represents the sampled waveform. Call this array of voltage values g. After the points of g have been entered into the workspace of Matlab, the Fourier components of g are found by typing

$$\texttt{G = fft(g).} \qquad (15.8)$$

The transform always has the same number of elements as the original sampled time waveform, so the complex vector G is 16 elements long. [Incidentally, the original time waveform can be reconstructed by performing an *inverse* discrete Fourier transform on G, using the Matlab function $\texttt{ifft}$. Type

$$\texttt{g = ifft(G)} \qquad (15.9)$$

and g will be identical to the original time waveform.]

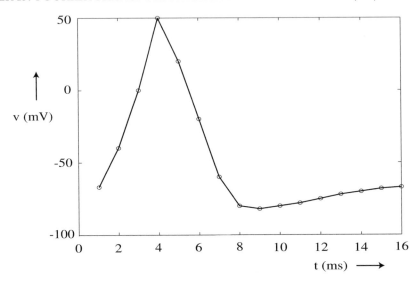

Figure 15.5: Simplified model of an action potential voltage waveform. This action potential repeats every 16 ms. The vector of sample points (denoted by circles) is labeled g.

Although the frequency spectrum G is readily found in Matlab by the command (15.8), some care must be taken in interpreting the results. In particular, *four* issues must be dealt with in interpreting the frequency components given by G:

1. *Converting to real values* – The vector G is *complex*, meaning that each element of G [for example, $G(i)$] has two parts: a real value and an imaginary value. Complex values are used here because the FFT algorithm is more efficient when put in terms of complex exponentials. For our use, we need to convert the complex notation of G into an all-real form (amplitude and phase) that can be used in formulas like (15.6). To do this, we first find the amplitude A of the frequency components by taking the absolute value of each element of G, multiplying by a factor of 2, and dividing by the number of samples N. In Matlab we enter

$$A = 2*abs(G)/N. \tag{15.10}$$

The phase of the components is the inverse tangent of the ratio of the imaginary to real parts of G, found in Matlab by typing

$$phi = angle(G)*180/pi. \tag{15.11}$$

The factor 180/pi in (15.11) converts the angle from radians to degrees. A and phi are real vectors 16 elements long. Plots of A and phi for the waveform above are shown in Fig. 15.6.

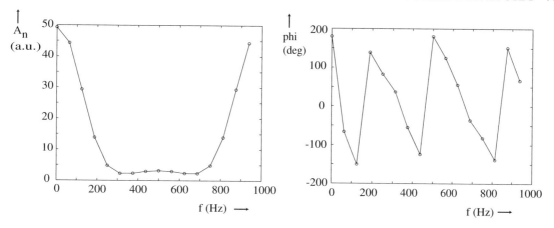

Figure 15.6: The amplitude and phase of the frequency components of the action potential shown in Fig. 15.5, before the redundancy is removed.

2. *Removing redundancy in A and phi* – Because the time waveform we started with in Fig. 15.5 is real (as are nearly all of the waveforms of interest in practical signals), the Fourier transform is **Hermitian.** That means that the right half of both the A and phi vectors in Fig. 15.6 are mirror images of the left half, and all necessary information is contained in the first half (9 elements, including the first, or dc, term) of each[3]. Thus, without loss of information we can cut the 16-element A and phi vectors in Fig. 15.6 nearly in half, preserving only the first 9 elements. The results are shown in Fig. 15.7.

3. *Special treatment of the dc term* – In Step 2 above we removed the right half of the spectrum representing the redundant negative frequencies. Since $\cos(\alpha) = \cos(-\alpha)$, these negative frequencies were compensated for by using the factor of 2 in the multiplication of (15.10). But the amplitude of the dc component is given by the first element of A, which is $A(1)$. This dc term, having zero frequency, should *not* be multiplied by 2 since it is not included as part of the negative frequencies. Instead we should multiply by a factor of 1. Thus, $A(1)$ is calculated in Matlab by typing

$$A(1) = abs(G(1))/N. \tag{15.12}$$

Also note that the dc term (i.e., for $f = 0$) is located in the first element $(i = 1)$ of A, not the "zeroeth" element, since Matlab starts with the index $i = 1$; it does not allow an index $i = 0$.

4. *Determining the frequency scale* – The last step is determining the correct scale for the frequencies of the Fourier components. The vector g describing the original time waveform does *not* contain any information about the actual time scale. That is, it is merely an indexed array of sampled values: $g(1), g(2), g(3)$, etc. It is up to the user to know that the time interval Δt between the samples has a

---

[3]The right half of the complex vector G actually represents negative frequencies, needed in the general case of a complex time signal. But since we are dealing here only with real signals, the right half of G (and A and phi) is redundant with the left half.

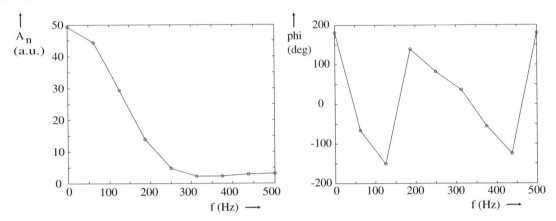

Figure 15.7: The amplitude and phase after the redundant negative-frequency right half has been re-moved.

certain value. Similarly, the frequency spectrum A does not contain any indication of the frequency scale; it is an indexed array of spectral amplitudes: $A(1)$, $A(2)$, $A(3)$, etc. *Therefore it is up to the user to calculate the frequency scale.* This is done as follows:

There are N samples in the time vector, each Δt apart. The total length of the time sample is therefore

$$T = N\Delta t. \tag{15.13}$$

(It is implicit that the time signal is repetitive, so T is also the period between repetitions of the signal.) The frequency spectrum from the FFT algorithm also has N samples. The frequency spacing between each sample Δf is given by

$$\Delta f = 1/T, \tag{15.14}$$

and therefore the total frequency length of the spectrum of g is, using (15.13),

$$F = N\Delta f = N/T = 1/\Delta t. \tag{15.15}$$

But after the vectors A and phi are cut in half (Step 2) to eliminate redundancy, the length of the frequency spectrum is

$$F_0 = F/2 = 1/(2\Delta t). \tag{15.16}$$

Notice from (15.16) that the highest frequency possible in the calculated spectrum is $f_h = F_0 = 1/(2\Delta t)$, which is why the Nyquist criterion (15.7) was used to determine the minimum original sampling rate.

Figure 15.7 above was obtained by applying these four steps. It plots the amplitude A and the phase phi of the frequency components of the action potential in Fig. 15.5, found by using (15.8), (15.10), (15.11), and (15.12). The correct frequency scale in Fig. 15.7 was found using the following calculations:

From Fig. 15.5, $\Delta t = 1$ ms and $N = 16$, so from (15.13), $T = 16$ ms.

Then from (15.14), $\Delta f = 1/(16 \text{ ms}) = 62.5$ Hz,

and from (15.16), $F_0 = 1/(2)(1 \text{ ms}) = 500$ Hz.

Example 15.1. Finding a Frequency Spectrum

The voltage recorded by a small electrode placed inside a certain sensing cell is repetitive. One cycle of this voltage waveform is shown in Fig. 15.8.

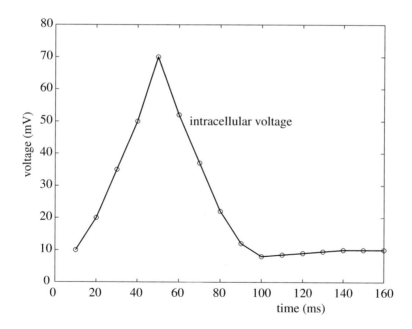

Figure 15.8: Sample voltage waveform whose spectrum is found in Example 15.1.

Tasks

a. Find the fundamental frequency of this waveform.

b. Use the FFT function in Matlab to find and plot the amplitude and phase of the Fourier frequency components of the waveform.

Solution
a. From (15.4), $f_1 = 1/T = 1/160$ ms $= 6.25$ Hz.
b. First we enter in the command window of Matlab a 16-element vector g of equally spaced samples of the voltage (obtaining the value at each sample point from the graph above), and the number of samples N:

```
g = [10 20 35 50 70 52 37 22 12 8 8.5 9 9.5 10 10 10];
N = 16;
```

Then we find the complex Fourier spectrum of g using (15.8):

```
G = fft(g);
```

To calculate the real amplitude A and phase phi of the frequency components, we follow Steps 1–3 above using (15.10), (15.11), and (15.12). Therefore, concerning ourselves with only the first half (9 elements) of A and phi, we enter

```
A = 2*abs(G(1:9))/N;
phi = angle(G(1:9))*180/pi;
A(1) = abs(G(1))/N;
```

To plot A and phi, we need to find the correct frequency scale (Step 4). Since $T = 160$ ms and $\Delta t = (160 \text{ ms})/16 = 10$ ms from the original waveform, (15.14) gives $\Delta f = 1/160$ ms $= 6.25$ Hz, and (15.16) gives the length of the (half) spectrum as $F_0 = 1/(2 \times 10 \text{ ms}) = 50$ Hz. So we set up the frequency scale by entering

```
f = [0:6.25:50];
```

and plot A and phi by entering

```
plot(f,A,f,A,'o');
xlabel('frequency (Hz)');
ylabel('amplitude A');
figure
plot(f,phi,f,phi,'o');
xlabel('frequency (Hz)');
ylabel('phase (deg)');
```

The results are shown in Fig. 15.9.

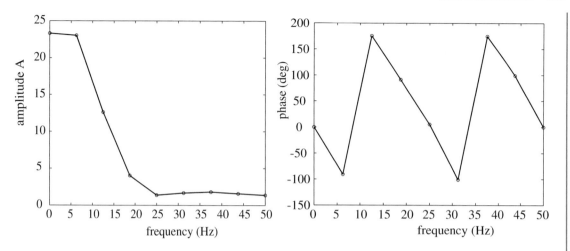

Figure 15.9: Spectrum of sample waveform in Example 15.1.

15.4   PROBLEMS

15.1. m. An electrocardiogram (EKG) recording for a healthy heart has the approximate waveform shown in Fig. 15.10.

It repeats every 800 ms for a resting heart.

a. Calculate the fundamental frequency of this waveform.

$$[\text{ans: } f_1 = 1.25 \text{ Hz}]$$

b. Using Matlab, find the amplitude A and the phase Φ of the frequency components of this wave. Sample the waveform at 16 uniformly spaced points. Then plot the amplitude A and phase Φ as a function of frequency. You only have to turn in these plots as your homework answer. Be sure your frequency scale is correct!

$$[\text{ans: see plots below}]$$

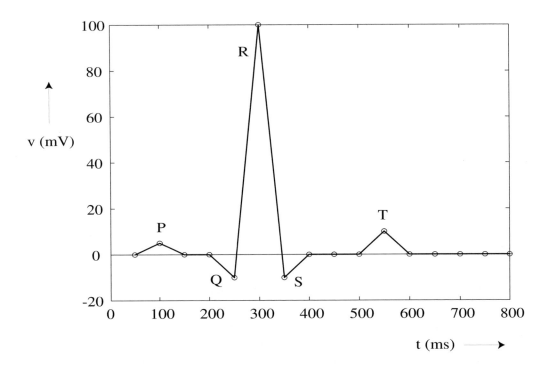

Figure 15.10: EKG waveform analyzed in Problem 15.1.

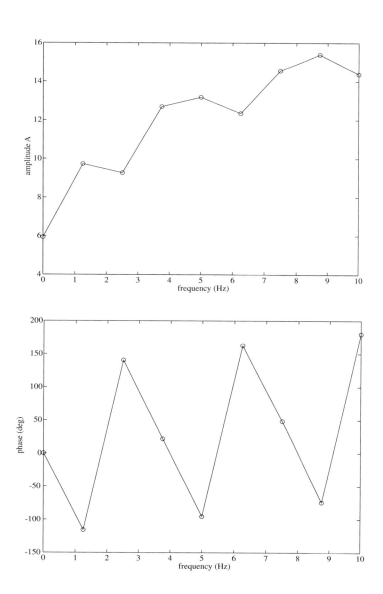

Figure 15.11: Spectrum results of Problem 15.1.

APPENDIX C

Major Project

Note on Organization of Major Project

This project can be arranged into two halves:

Computer modeling only (1/2 semester) – Section C.1 (Background) and Section C.2 (Matlab Model).

Electrical circuit modeling only (1/2 semester) – Section C.1 (Background) and Section C.3 (Electrical Circuit Analog).

For a full-semester project, all sections are covered.

C.1 BACKGROUND (AND PRELIMINARIES)

C.1.1 MODELING THE CARDIOVASCULAR SYSTEM

Models are used extensively in engineering analyses. This is sometimes because experimenting with the real thing is too expensive or cumbersome (e.g., determining the failure modes of a large bridge during an earthquake, or finding the most likely fracture planes of a hip-joint replacement fixture), and sometimes because experimenting would be unethical (testing chemical warfare agents in public places, or testing the effects of zero blood flow to the kidneys in humans). In such cases, computer models, animal models or electrical models can be used to simulate the procedure and predict its various effects.

NASA is interested in finding how the human cardiovascular (CV) system behaves under the unusual environmental conditions that astronauts will encounter. These include the high acceleration forces (up to 4 G) that astronauts experience during lift-off, and the zero gravity (0 G) environment they encounter in orbit and outside the earth's gravity field. Gravity plays an important role in determining the distribution of blood volume in various parts of the body, so changing gravity conditions (especially when protracted) can alter the functioning of the entire CV system.

To study the effects of various gravity conditions on the heart, vessels, blood flow, and blood pooling, NASA relies upon computer models. For example, a study at the University of Utah led by Dr. Keith Sharp (then a research associate professor in the Department of Bioengineering) and performed by graduate student Kristy Peterson developed an extensive computer model of the human CV system[1]. This research predicted the changes in blood distribution for several posture positions that the astronauts might take during lift-off and under zero-gravity exposure. Our computer model

---

[1]K. Peterson, *A Numerical Simulation of the Cardiovascular System to Investigate Changes in Posture and Gravitational Acceleration*, Master of Science Thesis, University of Utah, Dec. 1999.

for this major project will be similar to their model (though much simpler), but without any gravity effects. However, it will still give valuable insight into the behavior of the human circulation system for both healthy and diseased physiology.

C.1.2   OVERVIEW OF MAJOR PROJECT

A major goal of this project is to use engineering modeling and simulation tools for examining the dynamic behavior of the human CV system, which is shown in a simplified anatomical drawing in Fig. C.1. Although we will focus on the CV system, the techniques you will learn can be applied to a vast variety of other bioengineering problems, such as orthopedic modeling, imaging, biosensing, and biomaterials behavior.

We will use a lumped-element model of the CV systemic system, explained in much more detail in Section C.1.5. It has been simplified in many regards, such as ignoring the pulmonary circulation and using only a single-chambered heart in order to be more easily analyzed, but it still retains enough detail to display and predict many of the essential features of the CV system. It is essential in our model to keep the two valves that form the entrance port (the mitral valve) and exit port (the aortic valve) of the left ventricle. The arrangement of these two valves around the left ventricle is depicted near the top in Fig. C.1.

The body of this project is comprised of two major sections:

Matlab (Section C.2) – In this part, you write equations relating the variables of the CV model to each other, then numerically solve these equations using the Matlab computer language by stepping through time in small increments during a heart cycle. As input parameters to the Matlab program, you will need to calculate typical values for a healthy CV system. You will plot pressures and other essential quantities during successive heart cycles for a healthy circulation. Then you will model three disease states to see their effects on CV performance.

Electrical Analog (Section C.3) – In this part, you will see that electrical circuits follow the same equations as fluid-mechanical circuits (such as the CV system), once pairings that relate electrical quantities to corresponding fluid quantities are made. The electrical circuit is identical to the Matlab model, except that a special op amp circuit is needed to model the left ventricle and an extra branch is needed to add or bleed "blood volume" to the circuit. You will determine the values of the circuit components, then assemble the circuit, run it, and measure equivalent CV variables (such as the voltage waveform representing the aortic blood pressure waveform). You will initially model a healthy heart, then change the circuit to represent two more disease states to see their effects on CV performance.

C.1.3   NOTEBOOK REQUIREMENTS

One hallmark of a good engineer is that she or he always keeps complete and accurate records of derivations and experimental observations. The record should be so complete that another engineer can follow the derivation or repeat the experiment several years later based upon this written record

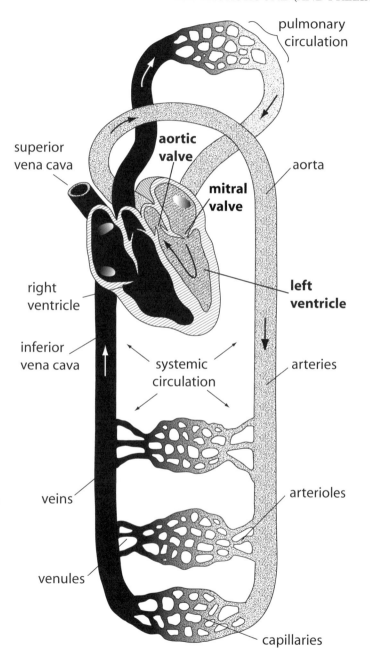

Figure C.1: A much simplified schematic diagram of the human circulatory system, showing major categories of vessels. Our heart model will use only the left ventricle, its two valves—the mitral (entrance) valve and the aortic (exit) valve—and the systemic vessels. Thus, the pulmonary circulation will not be included in our model. (Adapted from Guyton and Hall, Textbook of Medical Physiology, Saunders, 2000, Fig. 14-1.)

alone. This means that it needs to contain all the detail that was originally needed to solve the problem, including even false starts and mistakes!

Another reason for keeping a complete record is for patent purposes. Many organizations have a "patent notebook" policy requiring engineers to keep a legal record of all work as proof of inventorship. This involves keeping a contemporary record (on-the-spot entries) in a bound notebook, in ink, with all pages numbered, dated, and signed. For this project, you will gain experience keeping this type of notebook. All written work related to the project must be recorded in this notebook. The **rules** for keeping the notebook are:

1. The notebook must be about 8 1/2 x 11" in size with *fastened-in pages*; no folders or loose-leaf binders. This means that no pages can be added or taken away from the original. A spiral-bound notebook will suffice for us. (Actually, true patent notebooks should have sewn-in pages, but these are somewhat expensive for us.)

2. *All* entries must be in ink. If you make a mistake, cross it out with an X and start again. Do not try to erase or obliterate.

3. All pages must be numbered in order from the front of the notebook. Do this in ink by hand. You can use one or both sides of the page (use one side if your writing shows through the paper).

4. Make all your entries immediately in the notebook. Do *not* do them first on a loose piece of paper to transfer later—you won't have time. Your notebook does not have to be overly neat, just complete and readable by someone else like a fellow engineer or the teaching assistant. Don't spend a lot of time making the tables and drawings fancy, just complete and clear.

5. For purposes here, you can tape in computer printouts and graph printouts on blank pages of the notebook to keep a record of all your work.

6. After you have filled a page, date it in ink at the top and sign each page.

C.1.4   GRADING AND CHECKOFF DATES

There are **four checkoff periods**: two for the Matlab section and two for the electrical circuit section. You will be required to show your work and resulting data to the teaching assistant four times during the semester in your scheduled lab session within the weeks shown in the schedule given in the class syllabus. Your grade on the major project will be determined by how well you have followed and completed the tasks given in this handout, as evidenced by your notebook entries and records, your filled-in tables, your printouts, your assembled circuit, measurements on your circuit, and answers to questions by the teaching assistant.

No checkoffs of any Major Project section will be allowed after the scheduled week for that section. If you have any questions about how you are graded, please first see the teaching assistant, then the instructor if you still have unresolved questions. Table C.1 shows the four checkoff sections, what items will be graded, and how many points are assigned to each item.

	Checkoff	To Be Completed	Points
Matlab Model	#1	❏ Good notebook practice (guidelines followed). ❏ Filled-in Table 2. ❏ Filled-in Table 4. ❏ Calculations of all fluid R & C values (in notebook). ❏ Equations relating CV variables (in notebook). ❏ Vectorized and Euler's form of these equations (in notebook). ❏ Matlab m-file with FOR loops **started** (display or printout) – *[Note: At this stage, the program does not have to be finished, just started.]*	15 15

	#2	❏ Matlab m-file completed (printout in notebook).
		❏ Matlab simulation for healthy CV parameters:
		   demo of program to teaching assistant,
		   printout of graph of pressure waveforms (in notebook),
		   printout of parameter file for healthy CV (in notebook).
		❏ Matlab simulations of three diseases; for each disease:
		   demo of program to teaching assistant,
		   printout of graph of pressure waveforms (in notebook),
		   printout of parameter file for each disease (in notebook).
		❏ Relevant boxes in Table 5 filled in.
			15
			
			20

Electrical Circuit Model	#3	❏ Filled-in Table 6.
		❏ Calculations of all electrical R & C values (in notebook).
		❏ Circuit diagram for left ventricular module (in notebook).
		❏ Assembly of left ventricle module, including two voltage followers:
		   demo of working circuit to teaching assistant,
		   dual plot—by hand—of voltage waveforms for both excitation and v_h, with vertical axis in units of both V and equiv mmHg (in notebook). *[Note: points will be deducted if graph is not done carefully and completely.]*
			15
			
			20

	#4	❏ Assembly of entire systemic CV circuit.
		❏ Measurements of electrical waveforms for healthy CV system:
		   demo of circuit to teaching assistant,
		   dual plot of voltage waveforms for v_h and v_o (in notebook)—see comments above regarding grading of dual waveform plot.
		   record of cardiac output (in notebook).
		❏ Measurements of affected electrical variables for two diseases:
		   demo of circuit to teaching assistant,
		   graphs of voltage waveforms for v_h and v_o (in notebook),
		   record of cardiac output, percentage change, and resulting BP for each disease (in notebook).
		❏ Table 5 completed.
			20
			
			15

Table C.1:

C.1.5   OUR MODEL

A schematic of the CV model we will use for both the Matlab simulation and the electrical circuit simulation is given in Fig. C.2. It is obvious that several simplifying approximations have been made, as can be seen by comparing our model to even a simple anatomical drawing of the actual human CV system. But our model does have enough components to give the correct overall blood flow and pressure behavior, and to predict the effects of many of the common CV disease states, such as the high blood pressure that results from atherosclerosis.

Your first task will be to calculate from measured data of the healthy human CV system the values of the components in Fig. C.2, namely the resistance elements (R) that represent the pressure drop across major vessels and the capacitance elements (C) that represent the compliance of these windkessel vessels. Note in Fig. C.2 that we have lumped major categories of vessels together (for example, large and small arteries and arterioles) in the model for simplicity. This lumped-element approximation loses some information about spatial details, but helps keep the number of equations to a manageable level. The labels in Fig. C.2 show which elements correspond to the various segments of the CV system. With some study of the model, you will see that once a certain blood volume is inside the system, it just keeps circulating around the loop with no leakage out (unless some is bled away or added on purpose).

Also note in Fig. C.2 that in addition to the model parameters R and C, there are several variables which represent the state of the system at any given time. These include the pressure P of various vessels, the volume V in these vessels, and the volumetric flow rate Q through the vessels. These are the variables you will solve for using Matlab during time increments of a heart cycle. It is very important to keep track of the traditional clinical units used when measuring these variables and to *consistently* use these same units throughout the problem. To help keep track, fill in Table C.2 below with the clinical units you will use for each variable. (**Note:** The entire class should decide on common units to make communication and cross-checking easier.)

Table C.2: Traditional Clinical Units to be Used in Solution	
Variable	Clinical Units
Pressure P	
Volume V	
Volumetric Flow Q	

To keep the number of variables reasonable, we have modeled the heart by only one chamber—the left ventricle—which is the major pump responsible for ejecting blood into the aorta. Its pumping action is modeled by a *time-varying compliance*; the ventricle's compliance changes from a higher value C_{hd} during diastole to a lower value C_{hs} during systole in an exponential manner, as shown in Fig. C.3. This exponential behavior mimics reasonably well the compliance changes during an actual cardiac cycle, and is similar to the compliance waveform that will be generated by the electrical circuit in the second half of the project.

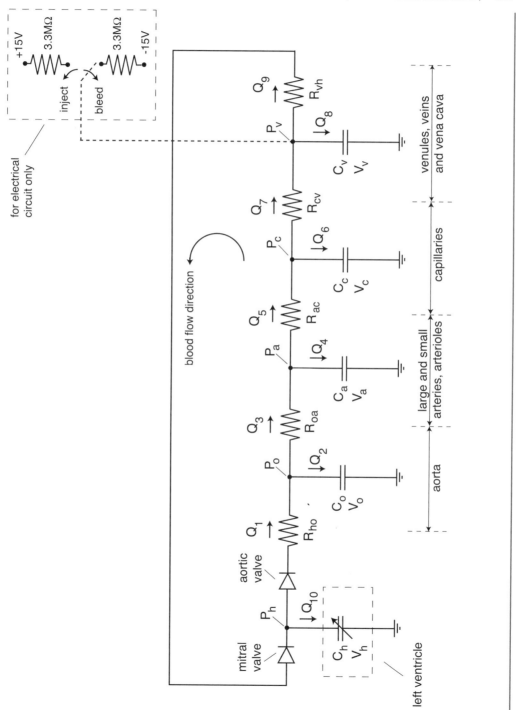

Figure C.2: Simple model of cardiovascular system.

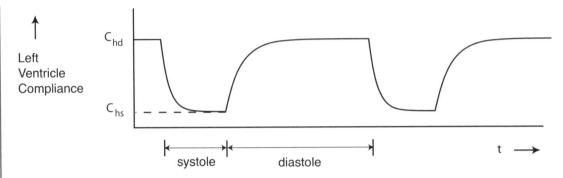

Figure C.3: Exponential model for the left ventricular compliance.

During the systolic segment, the ventricular compliance follows the equation[2]

$$C_h = (C_{hd} - C_{hs}) \; e^{-t/\tau_s} + C_{hs}, \tag{1}$$

where τ_s is the time constant representing the onset of systole ($\tau_s = 30$ ms fits our model best). During the diastolic segment, the compliance follows the equation

$$C_h = (C_{hs} - C_{hd}) \; e^{-t/\tau_d} + C_{hd}, \tag{2}$$

where τ_d is the time constant representing the onset of diastole ($\tau_d = 60$ ms fits best here).

To help you determine the compliance (C) and resistance (R) values of each of the vessel groups shown in Fig. C.2, you will need some important properties of a typical healthy human CV system given in Table C.3, including the blood volume contained in each compliant vessel category as well as the (estimated) residual volume in each compartment. Note that the total blood volume in the systemic loop (not including the pulmonary circulation) is 5.0 L − 0.8 L = 4.2 L. In addition, Fig. C.4 shows typical pressures around the systemic circulation, and Fig. C.5 shows the pressure waveforms of the left ventricle and aorta during one cardiac cycle. You will use these data sources to calculate the values of the model parameters in Section C.7.

C.1.6 APPROXIMATIONS

Any engineering model (mathematical, electrical or otherwise) must make some simplifying approximations to be useful; otherwise, it is so complex that it is impossible to solve. But these approximations can't go too far or the essence of the problem will be lost. For example, in our CV model we have reduced the four-chambered heart to a single chamber to keep the number of equations smaller, but we have to keep at least one chamber lest there would be no pumping action in the heart.

---

[2] In these equations, the time variable t is referenced to zero at the beginning of each segment. In the actual computer program, each segment must be shifted to match the systole and diastole timing shown in Fig. C.5.

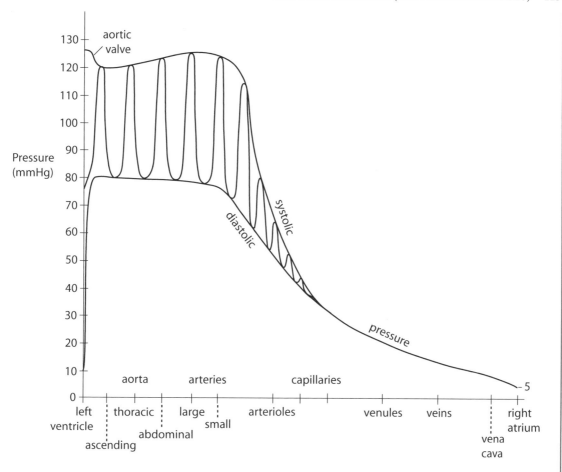

Figure C.4: Pressures around the human systemic circulating loop. Adapted from Enderle, Blanchard and Bronzino, Introduction to Biomedical Engineering, Academic Press, 2000, Fig. 10.7.

Some approximations have a minor effect on the accuracy of the model's predictions; others are more profound. While you are working on this project, fill in Table C.4 (pencil is okay here) with a list of *five* of the several approximations that we make, give a judgment as to whether each approximation has a minor, medium or major effect on the accuracy of the model, and list one thing **lost** by making each approximation. Fill in Table C.4 in time for Checkoff #1.

C.1.7 CALCULATION OF R AND C VALUES

This is the important first step in obtaining the parameters for our CV models. To do this, calculate values (from the data sources in this appendix for the healthy circulation) for all the elements (R's and C's) of the model of Fig. C.2, including the two C's that model the left ventricle, C_{hd} and C_{hs}. When

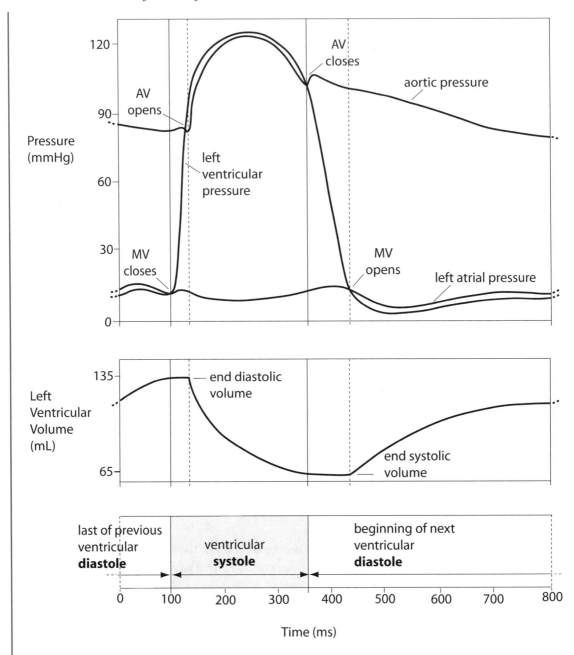

Figure C.5: Wiggers diagram showing the relationship between various blood pressures and ventricular blood volume during a cardiac cycle. Adapted from D. Silverthorn, Human Physiology, an Integrated Approach, Prentice Hall, 1998, Fig. 14-27.

Table C.3: Characteristics of typical human circulatory system			
Vessels	Contained volume (mL)	Residual volume (est.) (mL)	Average velocity (cm/s)
Left ventricle	100	0*	
Aorta	150	85	40
Arteries	350	170	10-40
Arterioles	50	20	0.1-10
Capillaries	300	60	<0.1
Venules	300	60	<0.3
Veins	2500	500	0.3-5
Vena cava & left artrium	450	90	5-30
Pulmonary circulation	800		
Heart rate (beats/min)	60-80	Cardiac output (L/min)	5-6
Total blood volume (L)	5.0	Stroke volume (mL)	70
Adapted from Berger, Goldsmith and Lewis, Introduction to Bioengineering, Oxford Press, 2001, Table 3.1. * Estimate from H. Senzaki et al., *Circulation* **94**:2497-2505, 1996.			

calculating compliances, take the residual volumes listed in Table C.3 into account. (However, assume that the residual volume of the *left ventricle* is zero because this gives a more reliable calculation of the heart's compliance.)[3] When calculating the resistances in the two segments containing valves (R_{ho} and R_{vh}), take into account the fact that the flow through these valves is pulsatile—i.e., that the cardiac output passes through these two valves only during the periods of systole and diastole, respectively. Be sure to include proper *units* for each value, following the units listed in Table C.2. Do the calculations in your notebook, and then **make a table** in your notebook that summarizes all these values.

C.2   MATLAB MODEL

TASKS:

2.1. Write equations relating all of the variables (P, Q, and V) shown in the model of Fig. C.2 to each other. You should use pressure/volume/compliance relationships, Poiseuille's Law, the conservation of volume law, and equations that relate volume changes to flow. Some sets of equations will be simple algebra and some will contain a time-derivative term. Also, some

[3]This assumption has been validated by a human study, which found that the left ventricle behaves dynamically as though its residual volume was approximately zero. See H. Senzaki et al., Circulation **94**:2497-2505, 1996.

Table C.4:			
#	Approximation	Effect (minor, medium or major)	One Thing That is Lost in Model
(Example)	Pulmonary circulation loop is neglected	medium	Pulmonary edema (pooling of blood in lungs) can't be modeled
1			
2			
3			
4			
5			

equations will be conditional (with an IF statement) describing the flow through the valves. You should end up with about 22 final equations, not counting the equations that generate the waveform for C_h. Do all the mathematics in your notebook.

2.2. Break the period of one heart cycle into many increments, each with time length Δt. Then apply *Euler's method* to approximate by forward differences the time derivatives of those equations developed in Task 2.1 that contain such derivatives.

Then write *all* of the final equations of Task 2.1 in *vector* notation. That is, let all the variables (for example: P_a or V_a) be represented by row vectors, whose elements $P_a(i)$ or $V_a(i)$ are the values of the variable at each increment of time. For those equations that use Euler's method, solve for the value at the next increment of time $V_a(i + 1)$ in terms of the value at the current time step $V_a(i)$. This will allow your Matlab program to step through time during a heart cycle. Since you will be coding these equations in Matlab in the next task, collect them all (after you've finished putting them in vectorized format) on one or two pages of your notebook.

2.3. Write a Matlab script m-file that contains two FOR loops. The *inner* FOR loop (the major one) steps through the time increments during one heart cycle (say 800 steps) solving for the time behavior of all the variables (P's, Q's, and V's) shown in Fig. C.2. This uses the equations developed in Task 2.2. Also inside this loop will be some conditional sections using IF statements to handle the conditional flow through the valves.

The *outer* FOR loop repeats the heart cycle again and again for as many times as necessary to reach steady state (perhaps 10 or more cycles). Be sure to "glue" the start of each cycle to the end of the previous cycle by initializing all beginning volumes in the new cycle $V_a(1)$ to their values at the very end of the previous cycle $V_a(801)$.

Also, figure out a way to calculate the cardiac output (CO) of each cycle. CO is defined as the blood volume being pumped out of the left ventricle per minute (so its units are L/min), but it can be calculated on a beat-to-beat basis. You will write out the CO after each heart cycle in the next task, along with writing out the blood volume in the venous side and plotting all the pressure waveforms after each cycle.

Before entering into the outer FOR loop for the first time, your program must first read in all the parameters for your model (for example, the R and C values for a healthy circulation, all timing parameters, and the residual volumes) and the initial distribution of the blood volume in the various compliant vessels. (**Important**: The distribution of blood volume throughout the system will surely adjust during the simulation run, so exactly what the distribution is when you start is not very important. But the *total* blood volume will remain constant, so make sure you have the proper total starting volume, which is 4.2 L for the normal CV systemic system.) It is convenient to use a *separate* m-file which contains all those parameters for a healthy circulation and to read in this separate file by "inputting" its name (without the .m extension) at the beginning of the run of your main script file. Then other abnormal circulation states (e.g., hypovolemia) can have their own parameter files.

Also, before entering the outer FOR loop for the first time, you should calculate a row vector representing the left ventricular compliance waveform C_h. Use the equations found in Section 1.5, with each segment (systole and diastole) shifted appropriately to correspond to the timing found in Fig. C.5.

2.4. In your main script m-file, put the capability to **plot** out on the *same* graph on the screen the following waveforms updated after *each* heart cycle:

  • Left ventricular pressure.

  • Aortic pressure.

  • Arterial pressure.

  • Capillary pressure.

  • Venous pressure.

On the graph axes, put labels and correct units. To identify each pressure, use separate line styles and colors for each of the five lines. Also, use the TEXT command to write inside the graph the values of: a) CO for the heart cycle, and b) total venous blood volume (venules, veins and vena cava) at, say, the start of that heart cycle. Put these values in a text line (somewhere inside the graph) that looks like this:

CO = 5.3 L/min   Venous Vol = 3.3 L.

At the end of the plot section, put a PAUSE command [e.g., pause(0.5)] to allow time for the program to print the graphs on the computer screen before continuing the FOR loop. Print out a copy of the final version of your main m-file and tape it in your notebook.

2.5. Run your script m-file using *healthy* circulation parameters for several cycles, until steady state is reached. On the graph on the screen, watch the changes in the pressure waveforms as the total blood volume redistributes itself throughout the various compartments of the system. Watch for changes in CO and venous volume. After steady state is reached, stop your program and print out the last graph, representing the behavior of a healthy circulation, and tape it in your notebook along with a copy of the healthy parameter file. Be prepared to demonstrate this program and answer questions about it during Checkoff #2.

2.6. Several CV diseases have been observed and categorized in humans. A few of the more important ones are listed in Table C.5. In this task, you are to simulate three of them—**anaphylactic shock, left-heart failure** (sometimes called congestive heart failure), and **hypovolemia**—by using your Matlab program to see what effects these pathologies have on CV performance. To see how each disease state should be modeled, consult relevant books or explore the internet to learn about each disease. Then make new parameter m-files representing each disease state with its modified values. Run your Matlab program for each of these abnormalities, and note the major effects on certain pressures, CO and venous volume. Print out a representative graph to tape in your notebook along with its corresponding parameter file for *each* pathology. Then fill in (pencil is okay) the various columns in Table C.5 for these three diseases. (The last two diseases will be simulated with the electrical circuit, next.)

C.3   ELECTRICAL CIRCUIT ANALOG

As you learned earlier, there is a one-to-one correspondence between the differential equations describing major fluid-mechanical components and the equations describing electrical circuits once the electrical variables are identified which correspond to the respective fluid variables. Thus, the model shown in Fig. C.2 can be directly implemented as an electrical circuit, and pressures and flows can be measured as electrical quantities from this circuit. This is the overall goal of Section C.3 of this project.

TASKS:

3.1. Fill in Table C.6 with the units you used in Table C.2 for the fluid CV variables, then with the names and units of the *electrical* variables that correspond to the fluid variables.

3.2. Translate all of the fluid-element values that you previously calculated in Section C.1.7 for a healthy circulation (all R's and C's) into their respective electrical-element values. To find the translation scaling factor for the C's, use the fact that that the largest compliance value found in Section C.1.7 should correspond to the largest practical capacitance size available in the lab (100 μF). Then, to find the translation scaling factor for the R's, use the fact that *time* is the same in both systems (fluid and electrical); the result is that the scaling factor for the R's will depend on the scaling factor found for the C's. Be sure to include units in your scaling equations. Then make a table of all the *electrical* R and C values for Fig. C.2 for a healthy circulation, with proper units. Do this work and record your values in your notebook. (Note: Another scaling equation—relating fluid pressure to electrical voltage—will be left unspecified until later, when it can be adjusted to keep the voltages of the circuit within a reasonable range.)

3.3. The left ventricle poses a special challenge for the electrical-circuit model. The capacitance of capacitor C_h must change from a high value C_{hd} during diastole to a low value C_{hs} during systole *without* changing the electrical charge (which is analogous to ventricular blood volume) stored on the capacitor during the rapid changeover—this corresponds to isovolumic contraction of the ventricle.

An active circuit that can achieve this by using two op amps and an analog switch is shown in Fig. C.6. The portion containing the two op amps is called a "capacitance multiplier" circuit since the *effective* capacitance seen across the input terminals (marked by X-X) can be larger than the physical capacitance value C_1. It can be shown that, using the branch current method and an ideal op amp model, the effective capacitance is given by

Table C.5:

Diseases	Probable Cause of Disease	Patient Symptoms or Complaints	What Parameters Change in Model? % Change?	Results: CO (L/min) % Change in CO?	Results: LV Pressure (mmHg) sys/dias, Aortic BP (mmHg) sys/dias
Aortic valve stenosis (this example will be done in class)					/ /
Hypovolemia					/ /
Left heart failure (either systolic or diastolic)					/ /
Anaphylactic shock (other than hypovolemia)					/ /
Arteriosclerosis (atherosclerosis)				proportional to	/ /
Aortic valve regurgitation				proportional to	/ /

Table C.6: Fluid/Electrical Pairs (Note: pencil is okay here)

Fluid Quantity	Symbol	Fluid Units	Corresponding Electrical Quantity	Symbol	Electrical Units
Resistance element	R			R	
Compliance element	C			C	
Pressure	P			v	
Volume	V			q	
Volumetric flow	Q			i	

$$C_{\textit{eff}} = (1 + \frac{R_2}{R_1})C_1.$$

In Fig. C.6, the analog switch is in the down position (open) during systole and is up (closed) during diastole; the resulting changes in $C_{\textit{eff}}$ are used to model the desired changes in C_h. During diastole, for example, the circuit resistance $R_1 = 10$ kΩ, but during systole the circuit resistance R_1 is effectively *infinite* in value as far as the leftmost op amp is concerned. In your notebook, write two expressions for C_h (that is, $C_{\textit{eff}}$) in terms of C_1, R_1, and R_2: write one for C_{hs} (systole—switch open) and one for C_{hd} (diastole—switch closed). Then use these expressions and the values for C_{hs} and C_{hd} you derived in Task 3.2 above to derive the values for C_1 and R_2 needed to correctly model C_h. Record all these values in your notebook.

(Note: The 4.7 μF capacitor is added here only so the variation in C_h follows a more gradual exponential waveform rather than an abrupt change between systole and diastole. Ignore the presence of the 4.7 μF capacitor in your calculations above.)

3.4. Redraw the complete left ventricular module diagram of Fig. C.6 in your **notebook**. Show pin numbers on your diagram by referring to the pin-out diagrams of the op amps (Fig. C.7) and analog switch (Fig. C.8). Show the power connections to the op amp chip and the switch chip. Also show values for the capacitors and resistors to give the desired values for C_{hd} and C_{hs} as determined in Task 3.3.

3.5. In the lab, assemble the left ventricle circuit you designed in Task 3.4 using the left half of a prototyping board, jumper wires, op amp chip, analog switch chip, and a function generator. Connect a power supply to the circuit and adjust the function generator to represent the cardiac

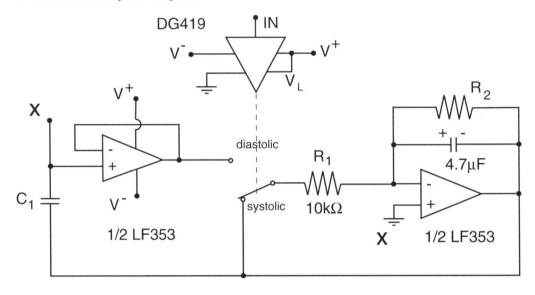

Figure C.6: Left ventricular module consisting of a capacitance multiplier circuit and an analog switch. The effective capacitance seen across terminals X-X represents C_h, which changes exponentially between a high value during diastole and a low value during systole.

cycle of Fig. C.5 with the proper systolic and diastolic periods. If you use electrolytic capacitors, be sure to observe the proper polarity of the leads by orienting the "+" lead toward the positive voltage and the "−" lead toward ground.

The voltage across the effective capacitance represents left ventricular pressure, and it will be a function of how much charge (representing blood volume) is stored on the capacitor. The scale factor relating circuit voltage to blood pressure can now be set. A good signal voltage range for the op amp is 0 to 4 V, so a recommended scale factor for translating between voltage and pressure is one such that **100 mmHg pressure is represented by a voltage of 4.0 V.** Thus if the oscilloscope vertical scale is set at 2 V per division, then a blood pressure of 100 mmHg is represented by a height of 2 divisions (2 "boxes") on the oscilloscope screen.

Run your left ventricle circuit and observe the changing voltage across the effective capacitor with an oscilloscope. To inject a proper amount of charge onto the capacitor in your circuit, use the charge injector arrangement shown in the dotted box in Fig. C.2. Inject enough charge into the capacitor to get a reasonable systolic blood pressure for a healthy left ventricle using the scale factor above. **By hand,** plot the voltage waveforms for both the excitation voltage (i.e., the voltage from the function generator driving the analog switch) and for the ventricular voltage v_h. Overlay the two waveforms on the same graph with the same horizontal axis (time) and the same vertical axis. Label the vertical axis with two scales: one in voltage (units of V) and one in equivalent blood pressure (units of mmHg) using the scale factor above. The graph

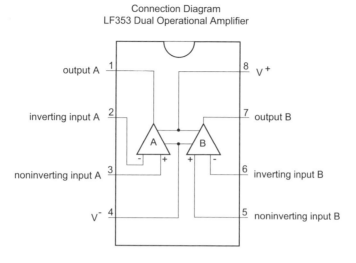

Figure C.7: Connection diagram for the LF353 op amps.

must be neat and complete, or points will be deducted from your score! Do the plot in your notebook, and be prepared to demonstrate this circuit during Checkout #3.

Important note: An oscilloscope has a large but *not* infinite input impedance, meaning that a small amount of current will be drained off the capacitor when the oscilloscope is connected, bleeding away most of the charge after a minute or two. This has undesirable consequences, both from an electrical point of view and a physiological point of view (obviously, excessive bleeding of a patient in order to take many repeated measurements of blood gases can lead to hypovolemia and eventual death). Therefore, we need to use an impedance "buffer" between the oscilloscope and the circuit to be measured. The arrangement in Fig. C.9 shows two **voltage followers**, and takes advantage of the extremely large (10^{12} Ω) input impedance of an op amp. Connect up two voltage followers on the top side of your prototyping board (using one additional op amp chip) between your circuit and the oscilloscope. Refer to the pin-out diagram of Fig. C.7 for correct connections. You can reuse the voltage follower setup that you have already employed in Problem 13.2. Connect one channel of the two-channel oscilloscope to the excitation (function generator) voltage and the other channel to the left ventricle (capacitor) voltage.

3.6. After the left ventricle module is working, you can assemble the entire circuit of Fig. C.2 on the prototyping board. Use the R and C values for a healthy heart found in Tasks 3.2 and 3.3. Inject enough charge into the circuit to get the correct systolic peak pressure in the *aorta* (this

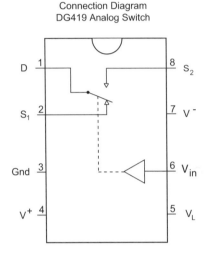

Switching Conditions

V_{in}	S_1	S_2
0 V	on	off
5 V	off	on

Figure C.8: Connection diagram for the DG419 analog switch.

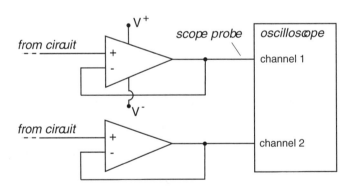

Figure C.9: Two voltage followers act as buffers between the electrical CV model circuit and the oscilloscope to avoid draining charge off the capacitors.

is close to the blood pressure measured in your upper arm) as given in Fig. C.5. There are several seconds of time lag between the time that charge is injected and the time the voltage settles to a final value, so have patience when injecting the charge. (The time lag represents the redistribution of blood within the various compartments of the body.) In order to measure the voltages v_h and v_o simultaneously, use the two voltage followers from the previous task connected to the channels of a two-channel oscilloscope. **By hand**, plot the voltage waveforms for both the ventricular voltage v_h and the aortic voltage v_o. Overlay the two waveforms on the same graph with the same horizontal axis (time) and the same vertical axis. Label the vertical axis with two scales: one in voltage (units of V) and one in equivalent blood pressure (units of mmHg) using the scale factor above. The graph must be neat and complete, or points will be deducted from your score! Do the plot in your notebook, and compare this graph to Fig. C.5. Be prepared to demonstrate your circuit performance and answer questions from the teaching assistant in Checkoff #4.

Since blood volume and electrical charge are analogs, cardiac output (CO) is represented by the average current flow around the circuit. To measure CO, you could insert an ammeter in any link of the circuit, but in practice the current is so low that most ammeters won't read it accurately. So use Ohm's Law to relate the voltage across one of the resistors to the current through it, and connect a hand-held voltmeter across this resistor to measure dc voltage. (R_{cv} is a convenient place to measure CO since the flow is rather steady here.) Record this voltage proportional to CO in your notebook after reaching steady state.

3.7. Now model the two remaining diseases in Table C.5, **atherosclerosis** and **aortic valve regurgitation**, using your electrical circuit. Determine which electrical components need to be changed (using only resistors) to represent each disease, then modify your circuit accordingly. For each disease, run the circuit, plot in your notebook the waveforms for v_h and v_o, record the new blood pressures in the ventricle and aorta, and record the new voltage proportional to CO. Calculate and record the *percentage* change in the CO compared to the healthy system. When adding, removing or changing resistors, keep the circuit running ("hot swap" the resistors) and avoid draining any charge off the capacitors while changing the resistors. Fill in the remaining boxes in Table C.5 based upon your results. You will be asked to demonstrate one of these pathologies during Checkoff #4.

GOOD LUCK (and remember to take care of your cardiovascular system during your lifetime)!

Bibliography

[1] Berger, S.A., Goldsmith, W. and Lewis, E.R. 1996. *Introduction to Bioengineering*, Oxford: Oxford University Press.

[2] Berne, R.M. and Levy, M.N. 1993. *Physiology*, St. Louis: Mosby-Year Book, Inc.

[3] Darcy, H. 1856. *Les Fountaines de la Ville de Dijon*, Dalmont, Paris.

[4] Durney, C.H. 1973. "Principles of Design and Analysis of Learning Systems," *Engineering Education*, March 1973, 406-409.

[5] Durney, C.H. and Christensen, D.A. 2000. *Basic Introduction to Bioelectromagnetics*, Boca Raton: CRC Press.

[6] Eide, A.R., et al. 1997. *Engineering Fundamentals and Problem Solving*, 3rd Edition, Boston: WCB McGraw-Hill.

[7] Enderle, J., Blanchard, S. and Bronzino, J. 2000. *Introduction to Biomedical Engineering*, San Diego: Academic Press.

[8] Folkow, B. and Neil, E. 1971. *Circulation*, New York: Oxford University Press.

[9] Fung, Y.C. 1984. *Biomechanics: Circulation*, New York: Springer-Verlag.

[10] Fung, Y.C. 1990. *Biomechanics: Motion, Flow, Stress, and Growth*, New York: Springer-Verlag.

[11] Guyton, A.C. 1974. *Function of the Human Body*, Philadelphia: Saunders.

[12] Guyton, A.C. and Hall, J.E. 2000. *Textbook of Medical Physiology*, Philadelphia: Saunders.

[13] Keener, K. and Sneyd, J. 1998. *Mathematical Physiology*, New York: Springer.

[14] Mars Climate Orbiter, on NASA web site: www.NASA.gov.

[15] Matlab Help Desk, The Math Works Inc., Natick, MA.

[16] Nillson, J.W. and Riedel, S.A. 1996. *Electric Circuits*, 5th Edition, Reading, MA: Addison-Wesley.

[17] Palm, W.J. III. 1999. *Matlab for Engineering Applications*, Boston: WCB McGraw-Hill.

[18] Peterson, K. 1999. *A Numerical Simulation of the Cardiovascular System to Investigate Changes in Posture and Gravitational Acceleration*, MS Thesis, University of Utah.

[19] Senzaki, H., Chen, C-H. and Kass, D.A. 1996. "Single-Beat Estimation of End-Systolic Pressure-Volume Relation in Humans," *Circulation*, 94: 2497-2505.

[20] Silverthorn, D. 1998. *Human Physiology, an Integrated Approach*, Upper Saddle River: Prentice-Hall.

[21] Starling, E.H. 1918. *The Linacre Lecture on the Law of the Heart,* London: Longmans, Green.

[22] Withers, P.C. 1992. *Comparative Animal Physiology*, Fort Worth: Saunders College Publishing.

Printed in the United States
by Baker & Taylor Publisher Services